T0328898

Energy Efficiency of Medical Devices and Healthcare Applications

Energy Efficiency of Medical Devices and Healthcare Applications

Edited by

Amr Mohamed, PhD

Professor
Computer Science and Engineering Department
College of Engineering
Qatar University
Doha, Qatar

Academic Press is an imprint of Elsevier
125 London Wall, London EC2Y 5AS, United Kingdom
525 B Street, Suite 1650, San Diego, CA 92101, United States
50 Hampshire Street, 5th Floor, Cambridge, MA 02139, United States
The Boulevard, Langford Lane, Kidlington, Oxford OX5 1GB, United Kingdom

Notices
Knowledge and best practice in this field are constantly changing. As new research and experience broaden our understanding, changes in research methods, professional practices, or medical treatment may become necessary.

Practitioners and researchers must always rely on their own experience and knowledge in evaluating and using any information, methods, compounds, or experiments described herein. In using such information or methods they should be mindful of their own safety and the safety of others, including parties for whom they have a professional responsibility.

To the fullest extent of the law, neither the Publisher nor the authors, contributors, or editors, assume any liability for any injury and/or damage to persons or property as a matter of products liability, negligence or otherwise, or from any use or operation of any methods, products, instructions, or ideas contained in the material herein.

Library of Congress Cataloging-in-Publication Data
A catalog record for this book is available from the Library of Congress

British Library Cataloguing-in-Publication Data
A catalogue record for this book is available from the British Library

ISBN: 978-0-12-819045-6

For information on all Academic Press publications visit our website at https://www.elsevier.com/books-and-journals

Publisher: Brian Romer
Acquisitions Editor: Zanol R
Editorial Project Manager: Leticia M. Lima
Production Project Manager: Kiruthika Govindaraju
Cover Designer: Vicky Pearson

Typeset by TNQ Technologies

Contents

Reviewers

- Lutfi samara,
  Department of Electrical Engineering, Qatar University, Doha, Qatar
- Alaa Awad Abdellatif,
  Department of Computer Science and Engineering, Qatar University, Doha, Qatar; Department of Electronics and Telecommunications, Politecnico di Torino, Torino, Italy
- Heena Rathore, PhD, BE,
  Department of Computer Science, University of Texas, San Antonio, TX, United States
- Abeer Al-Marridi, Ms in computing,
  Department of Computer Science and Engineering, Qatar University, Doha, Qatar
- Dr. Emna Baccour, PhD
  Postdoctoral fellow, Hamad Ben Khalifa University, Doha, Qatar
- Lamia Basyoni,
  PhD student, Qatar University, Doha, Qatar
- Naram Mhaisen,
  Master student, Qatar University, Doha, Qatar

Contributors

Abderrazak Abdaoui
Department of Computer Science and Engineering, Qatar University, Doha, Qatar

Alaa Awad Abdellatif
Department of Computer Science and Engineering, Qatar University, Doha, Qatar; Department of Electronics and Telecommunications, Politecnico di Torino, Torino, Italy

Sajjad Afrakhteh
School of Electrical Engineering, Iran University of Science and Technology, Narmak, Tehran, Iran

Abdulla Al-Ali
Department of Computer Science and Engineering, Qatar University, Doha, Qatar

Abeer Al-Marridi, Ms
Department of Computer Science and Engineering, Qatar University, Doha, Qatar

Carla Fabiana Chiasserini
Department of Electronics and Telecommunications, Politecnico di Torino, Torino, Italy

V. Divya, BE, MTech
Reseach Scholar, CSE, Thiagarajar College of Engineering, Madurai, Tamilnadu, India

Xiaojiang Du
Department of Computer and Information Sciences, Temple University, Philadelphia, PA, United States

Aiman Erbad, PhD
Department of Computer Science and Engineering, Qatar University, Doha, Qatar

Mohsen Guizani
Professor, Department of Computer Science and Engineering, Qatar University, Doha, Qatar

Ramy Hussein
Postdoctoral Fellow, Electrical and Computer Engineering, The University of British Columbia, Vancouver, BC, Canada

R. Leena Sri, PhD
Thiagarajar College of Engineering, Madurai, Tamilnadu, India

Amr Mohamed, PhD
Professor, Department of Computer Science and Engineering, Qatar University, Doha, Qatar

Mohammad Reza Mosavi
School of Electrical Engineering, Iran University of Science and Technology, Narmak, Tehran, Iran

Heena Rathore, PhD, BE
Department of Computer Science, University of Texas, San Antonio, TX, United States

Ali Riahi
Department of Computer Science and Engineering, Qatar University, Doha, Qatar

Rabab Ward
Professor Emeritus, Electrical and Computer Engineering, The University of British Columbia, Vancouver, BC, Canada

Preface

Continuous improvement of cost-effective healthcare and patient treatment is by far the top national interest worldwide. The proliferation, versatility, and agility of medical devices have revolutionized healthcare and contributed to the new Health 4.0 era of internet of medical things (IoMT). Broadly speaking, these medical devices can range from heavy equipment for patient treatment in medical facilitates, to miniaturized medical sensors, and finally implantable medical devices. Energy efficiency of such medical devices has enormous benefits, not only to provide cost-effective healthcare but also to improve patient care quality through continuous availability and robust measurements, which are key for the treatment of chronic diseases and emergency situations. Diverse characteristics of medical devices pose many challenges to provide energy-efficient healthcare systems, which in many cases require multidisciplinary approaches to provide comprehensive solutions.

Energy Efficiency of Medical Devices and Healthcare Applications book focuses on providing comprehensive coverage of cutting-edge and interdisciplinary research and commercial solutions in this field. Intended audience for this book include, but not limited to, academic and junior researchers, graduate students, technology developers/adopters, and those beginning a new line of research and development in this rich area. We discuss emerging technologies and issues in this field such as artificial intelligence (AI)-based, machine learning, and edge computing techniques to address various key design aspects of the medical devices and healthcare applications, such as energy efficiency, communications, hardware design, and security and privacy. We also discuss emerging application frameworks such as mobile/smart health and blockchain for health applications. We address energy-related trade-offs to maximize the medical devices availability, especially battery-operated ones, while providing immediate response and low latency communication in emergency situations, sustainability, and robustness for chronic disease treatment, in addition to high protection against cyberattacks that may threaten patients' lives. Finally, we discuss energy-related issues for enabling technologies and future trends of next generation healthcare, such as personalized health and IoMT, where patients can participate in their own treatment through innovative medical devices and software applications and tools. The book has a rich collection of carefully selected and reviewed manuscripts written by experts in the subject matter, discussing diverse technologies, case studies, and proposed novel techniques, while providing future directions for those who are interested to start a new line of research.

Amr Mohamed
Editor

CHAPTER

1

AI-based techniques on edge devices to optimize energy efficiency in m-Health applications

Abeer Al-Marridi, Ms [1], Amr Mohamed, PhD [2], Aiman Erbad, PhD [1]

[1] *Department of Computer Science and Engineering, Qatar University, Doha, Qatar;* [2] *Professor, Department of Computer Science and Engineering, Qatar University, Doha, Qatar*

1. Introduction

Healthcare is one of the highest priorities worldwide, where spending increases rapidly in this sector. In past years, the number of diseases increases rapidly, causing a vital rise in the number of patients compared with the number of doctors all over the world. The traditional way of communication between the patient and doctor cannot align with the situation. Owing to that, researchers consider the extensive use of mobile phones all over the world with the rapid development in technology domains, including smartphones, communication barriers, sensors, and much more, to support the shortage in health facilities.

The World Health Organization, defined that anything supports all the fields of healthcare through information and communication technology, goes under the electronic Health (e-Health) [1]. Mobile-Health (m-Health) is a subset of e-Health, which supports health objectives by deploying mobile telephone and wireless technologies [2,3].

The development of smart-phones devices raises new opportunities for researchers to integrate them into the treatment process. Therefore, smart-Health (s-Health) was defined as a component of m-Health. Smart devices eliminate the need for integrating separate sensors with the patients, as almost all these devices contain a built-in sensor for biosensing tracking [4]. Additionally, the connectivity problem will be eliminated using smart devices as the coverage of mobile cellular networks grows rapidly [2]. Fig. 1.1 is the Vann diagram that shows the relation between s-Health, m-Health, and e-Health.

2. Edge computing

In this chapter, edge computing will be discussed in general and mobile edge computing in specific. Edge computing considers a subset of cloud computing as it is the main aim to overcome cloud computing limitations. Mainly, it aims to

Energy Efficiency of Medical Devices and Healthcare Applications. https://doi.org/10.1016/B978-0-12-819045-6.00001-7

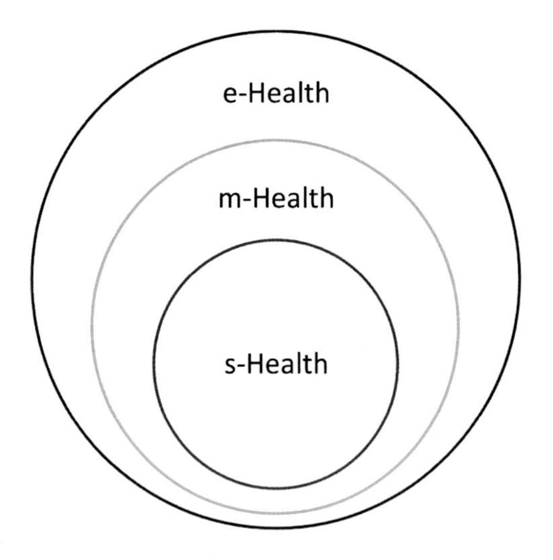

FIGURE 1.1

A Venn diagram shows the overlapping relationships, where s-Health is a subset of m-Health, which is, in turn, a subset of e-Health.

move the data computations and application away from the cloud servers closer to the end users to utilize the bandwidth and reduce latency.

The concept of edge computing eliminates the need for continuous communication with the cloud provider, where the interaction is mainly between the end users and the local servers to overcome the problem of high latency; however, it needs high bandwidth.

Edge computing makes use of the fast development of the devices in terms of technology, processing power, and batteries to reduce the movement and storage of data in the cloud to save both time and money. In m-Health systems, the deployment of edge computing is significant because it deals with a massive amount of medical data that needs to be processed and transmitted.

Many technologies and services were built based on the edge computing concept such as fog computing, mobile edge computing, cloudlet, wireless sensor networks, and much more, as shown in Fig. 1.2.

As mentioned before, edge computing came to overcome the limitations of cloud computing, such as the followings:

- Reducing the jitter and latency: Providing resources and services closer to the end users minimize the need for loading from the cloud centers. The locations of the servers in the edge computing are closers to the edge network, which is not the case with cloud computing. Reducing the latency means reducing the end-to-end delay, which is an essential requirement in many applications, especially the health-related application, as the delay may cause losing a patient's life.

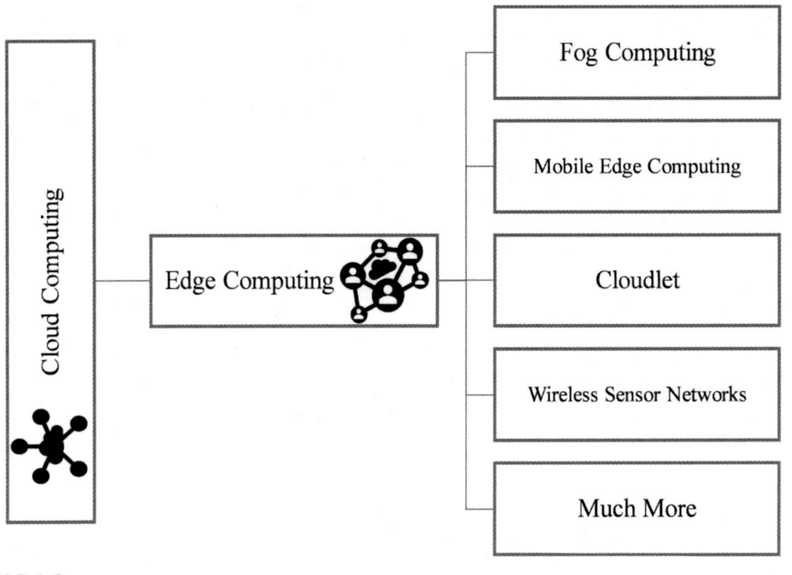

FIGURE 1.2

A flow chart showing the inheritance of edge computing technology from cloud computing to facilitate many other services and applications.

- Availability of the data: Having the computational resources and services closer to end users will benefit both the end users and the service providers. Service providers will have a clearer image of the needed resources and the allocation based on the behavior of the end users and their mobile user information.
- Supporting mobility and location-aware aspect: Supporting Locator ID Separation Protocol will facilitate direct communication with mobile devices. The location-awareness aspect in edge computing eases the property of accessing the closest server to their physical location and supports several edge computing applications.
- Severs distributions based on a dense, noncentric model to avoid a single point of failure.

At the same time, there are many challenges in integrating edge computing technologies, which encourage researchers to investigate more in the area and find different solutions. Such challenges like:

- Maintaining the security and privacy of the data as usual encryption techniques are not enough when dealing with edge computing architecture. Security techniques may add overhead latency to the communication process.
- Maintaining communication between heterogeneous devices, with different energy and performance constraints in a highly dynamic network with additional security requirements may affect the scalability of the network.

3. Data preprocessing on edge devices and transmission energy optimization

As mentioned earlier, due to recent trends and technologies such as the Internet of Things (IoT), the amount of processed, transmitted, and analyzed data increase rapidly, especially in e-Health systems. There are two types of transmitted data by a device, either raw data (unprocessed data) or context-aware data.

- The raw data are taken directly from the source without processing, which considered as meaningless if it is not understandable by the physicians.
- Context-aware (subject-oriented) is processed raw data that play a vital role in e-Health systems. Subject-oriented represents considering the correlation between the patient case and the detected data.
  - Example: The heart rate for a heart patient is different from the heart rate of ordinary people. Consequently, when the heart rate of heart patient reaches the regular heart rate of ordinary people in rest, then this should be considered as unusual behavior.

The preprocessing phase of that data is a significant phase of any smart system, especially when it comes to health. Due to that, researchers proposed and discussed different processing techniques in the literature to facilitate the usage of this vast data correctly in the healthcare sector.

Cleaning the data, dealing with missing values, removing outliers, and summarizing the data are examples of data preprocessing techniques that can be applied to the medical data to clarify its beneficial aspect to the receiver and increase the efficiency of m-Health systems.

Examples of data summarization techniques:

- Feature extraction
- Classification
- Protection
- Filtering
- Recognition detection
- Prediction
- Data reduction
- Adaptive compression

Examples of preprocessing algorithms used to accomplish different data preprocessing techniques:

- Bayesian network (BN)
- Conditional random Fields model (CRF)
- Machine learning (ML)
- Fuzzy systems
- Gaussian mixture model (GMM)
- Hidden Markovian models (HMM and its extensions)
- Deep learning (DL)
- Much more

Each algorithm has limitations and drawbacks that prevent it from being used in particular m-Health applications. For example, artificial neural networks (ANN) was employed in different m-Health systems as it used to have the best recognition and classification accuracy results in comparison with different approaches like support vector regression, decision trees, support vector machine, and logistic regression. However, the performance of the approach changes based on different factors related to the size and type of the medical data, types, and location of the sensors.

Various data processing approaches can be applied to the collected data to facilitate a particular objective and solve a specific problem. However, this might eliminate the benefit of getting extra information about the situation of the patient and prevent any future health failure.

As a result, different approaches were proposed to manage big data delivery using efficient data reduction techniques for analysis, and efficient transmission over bandwidth-constrained networks.

Data reduction is one technique to change the representation of the data. Many researchers proposed different methods that aim to reduce the data size [5–8]. Further to this, some researchers have taken into consideration the network and application requirements and constraints, such as [9]. The authors in Ref. [9] used fuzzy formal concept analysis with a smart sensing approach to optimize the complexity, reduce the stored data size, and maximize the lifetime of the battery-operated devices that expected to run without replacement for a long time. In Ref. [10], a modification of the block sparse Bayesian learning (BSBL-BO) method was performed to manage the dependency of received medical data. However, the computational complexity of the technique is high and cannot be deployed efficiently at battery-operated devices.

In this chapter, we will focus on compression as one type of data reduction methods. In compression, lossless and lossy compression are two types of data compression. Fig. 1.3 classifies different data reduction techniques under three classes.

- Lossless compression ensures reconstructing the original data from the compressed version without distortion. However, the compression ratio will be limited; as it would be hard to apply high compression ratios
  - Examples:
    - Run length encoding (RLE)
    - Arithmetic encoding (AE)
    - Lempel–Ziv–Welch
    - Entropy coding
    - Shannon coding
    - Huffman coding
    - Combination of DCT and ANN (near lossless)
- Lossy compression can achieve high compression ratios, but it introduces distortion on the reconstructed signal.
  - Examples:
    - Discrete wavelet transform (DWT)
    - Discrete cosine transforms (DCT)
    - Set partitioning in hierarchical trees
    - Compressive sensing

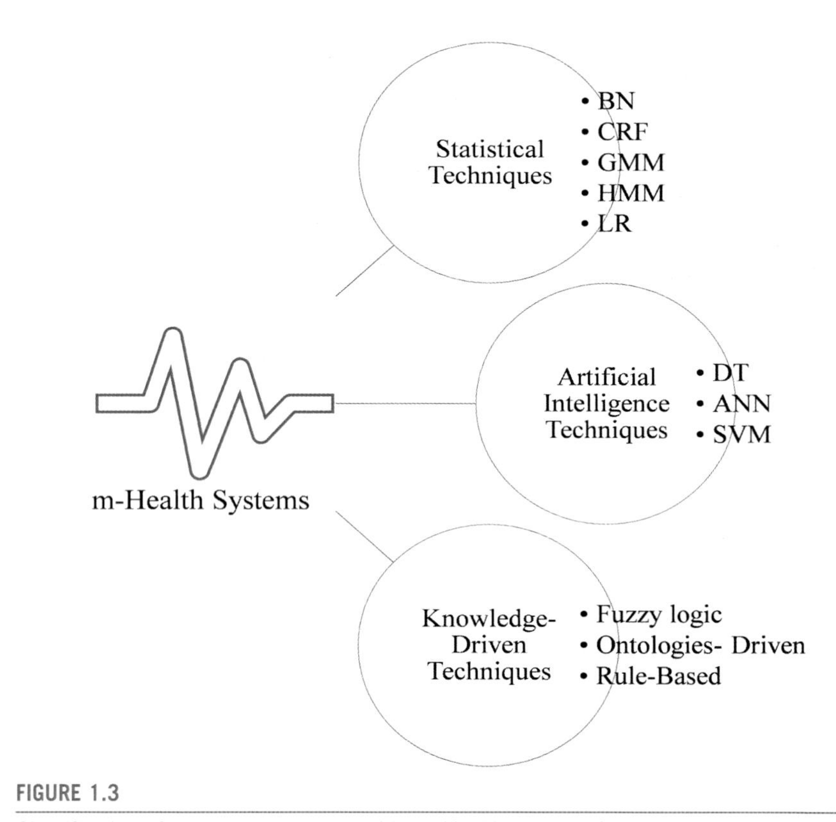

FIGURE 1.3

Classification of some techniques used in m-Health systems for data preprocessing.

- Hybrid compression
  - Examples:
    - Combination of DCT and Huffman coding.
    - Combination of either DCT or DWT and RLE or AE.

4. Deep learning for medical data preprocessing

Deep learning (DL) is a subfield of machine learning (ML), which aims to introduce artificial intelligence in machine learning. The intelligence learning in DL categorized into three fields, supervised, semisupervised, or unsupervised learning. A deep neural network is composed of multiple layers for feature extraction, where the output of a layer is used as an input for the successive layer, and each layer consists of a certain number of neurons, as shown in Fig. 1.4. Neurons work like the human brain to learn without the need for human input. The neurons in the first layer extract the simple features of the input and pass them to the next layer, which learns with time more detailed features about the input.

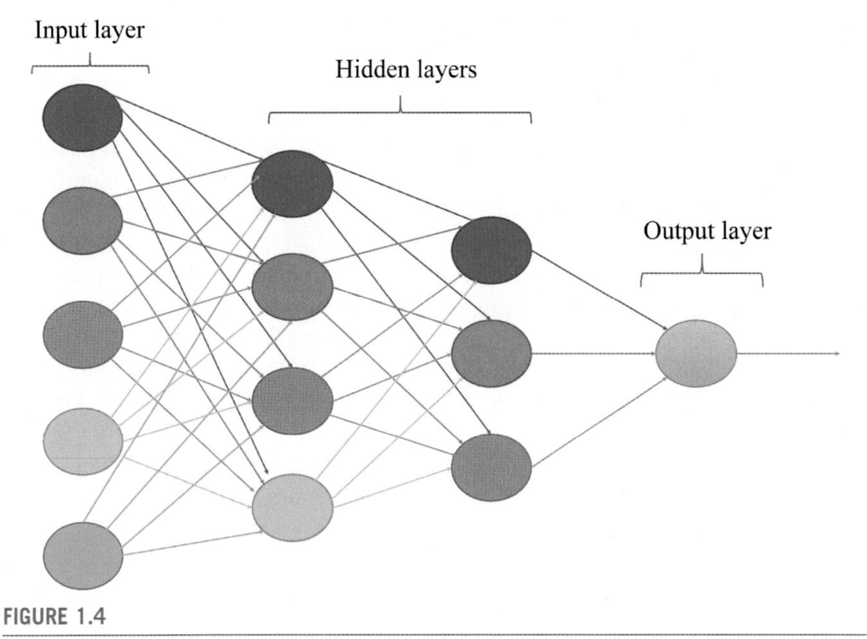

FIGURE 1.4

Simple representation of a neural network, where the circles correspond to neurons.

There are several neural networks architectures to solve different problems, such as the following:

- **Recurrent neural networks long/short-term memory**:
  Recurrent neural network (RNN) is used with the data that have good interdependencies to preserve information about what happened in all the previous layers, where the output of a layer depends on the previous computations. This type of neural network does not go through the update operation of the weights, as it uses the same weights across all the layers. Due to that, a reduction will occur in the total number of learned parameters. It is worth to mention that RNN has excellent achievements in language modeling, bioinformatics, and speech recognition applications [11].
- **Recursive neural networks**:
  Similar to RNNs, it has a shared weight matrix; however, it uses a variation of backpropagation called backpropagation through the structure as it uses binary tree structure in learning [12].
- **Convolutional neural networks**:
  Convolutional neural network (CNN) is usually used with data that include internal local correlations that need to be taken into consideration. CNN replaces the usage of neurons with filters that perform complex operations on the data. The density of connections between the layers in the network decreased by the usage of filters. Recent research shows that CNN outperforms the state-of-the-art

techniques in advanced natural language processing tasks and computer vision [13].

The main types of layers used in a CNN:

- Fully connected layer:
  They are usually used as the last layer after the convolutional and pooling layers to convert the 2D input into a 1D output. Nonlinear activation functions used in these layers to get the class prediction of the output [14].
- Convolutional layer:
  It consists of filters with precise dimensions to extract the features from the input. Filters are another representation of neurons, which generate an output value of a weighted input. The filter should move through the input and capture the features. If the input size is not divisible by the filter size, then the padding technique is performed on the input. Convolutional layers can expect an input on either typical pixel values if it is an input layer or feature map if it is a hidden layer. Feature map is the output of a filter in the previous layer, where we might get multiple feature maps from the previous layer based on the number of filters used [14].
- Pooling layer:
  It is called the downsampling layer, which is usually located after the convolutional layer to reduce the spatial dimensions of the input for the next convolutional layers. Pooling layers are usually used to minimize the chance of getting overfit, computational overhead, and cost.

- **Autoencoder**
  AE semisupervised learning approach aims to recreate the input from the output rather than classify it under a specific class. AE encodes the data with lower dimensionality to extract the discriminative features. The general representation of AE is depicted in Fig. 1.5.

Neural networks were used for compression along with other compression approaches. However, some researchers used different deep neural networks alone to compress medical data such as vital signs including electroencephalogram (EEG), electrocardiography, electrocorticographic (ECoG), and electromyogram. Stacked autoencoder was used by Ref. [15] for EEG compression, and the model was able to outperform the state-of-art methods at high compression ratios. However, the usage of convolutional layers instead of fully connected layers should enhance the performance because arranging the EEG data into a 2D formulation will consider the spatiotemporal correlation among the samples.

The rest of the chapter will discuss the efficiency of using convolutional autoencoder (CAE) in compressing sensitive medical data (e.g., EEG signs) and the integration of this approach in the edge devices (e.g., patient data aggregator, PDA) while taking into account both the network resources and m-Health system obligations.

In-network processing focuses on processing vital signs at the edge network **adaptively** to respond to wireless network dynamics while addressing

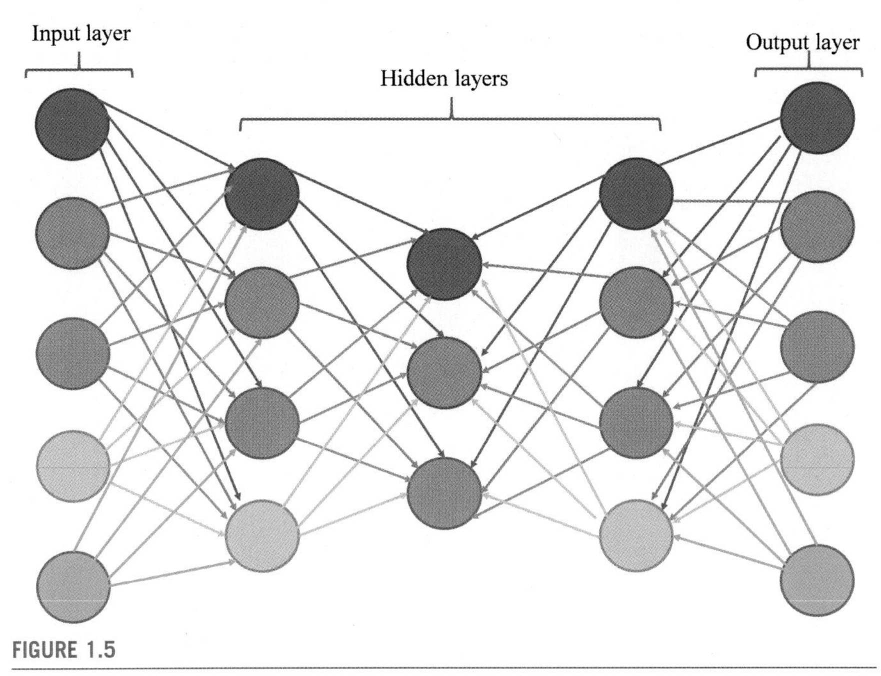

FIGURE 1.5

General representation of autoencoder.

application-level requirements. Data reduction plays an essential role in network resource allocation as optimizing the network and maximizing the use of resources is always a goal. Recently, processing and transmitting medical data through the network became a very active research area owing to the enormous enhancement that arises in wireless and mobile communication technologies. Many researchers went through different preprocessing, resource allocation schemes, and optimization frameworks to address the challenges behind the design of m-Health, such as the limitation of small sensors, which get affected by the consumed energy, storage, and computational resources, let alone addressing the application-level requirements.

An optimization cross-layer framework based on energy distortion with five decision parameters is discussed. The mobility of users is considered using the spatio-temporal parametric stepping (STEPS) model, where compressing and transmitting the data while the user is moving should be applicable, as it will affect some parameters of the optimization problem.

The output STEPS model is the location of all the users at each time slot of the simulation in a two-dimensional array with reference to the base station. The channel allocated for a user will be affected by the distance, causing changes in the transmission energy and the distortion.

Fig. 1.6 represents the overall system model where each patient connected with wearable devices that send vital signs signals to their edge device, "Patient Data

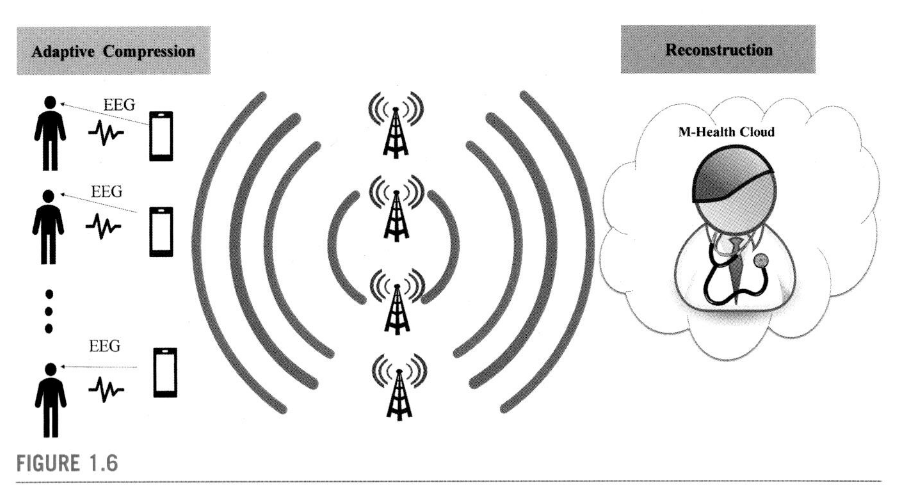

FIGURE 1.6

System model for adaptive compression at the edge.

Aggregator (PDA)." PDA aggregates the received data and runs the optimization to decide the appropriate compression percentage applied to the data using a convolutional encoder concerning the network and application obligations. The compressed data will be sent over the network to the m-Health server where the physicians apply the convolutional decoder to reconstruct the vital signs with the minimum accepted distortion rate for accurate and fast diagnosis of the patient case.

4.1 Compression using convolutional autoencoder

CAE is a combination of two architectures, the CNN and AE, as shown in Fig. 1.7. The encoder part of the network is considered as a compressor, which deployed on the edge device of the m-Health system. The decoder should reconstruct the compressed data with the minimum distortion rate by physicians at the receiver side of the m-Health system. The usage of CNN layers in the architecture enables the reconstruction of the compressed data with an acceptable distortion rate even at high compression ratios.

4.1.1 Data preparation

Preparing the data is the first step before applying compression using the convolutional encoder for two reasons: first, increase the overall efficiency of the system; second, apply high compression ratios smoothly at the edge devices with minimum reconstruction distortion at the m-Health servers. CAE approach is used on vital signs as medical data in specific EEG and ECoG data. Therefore, the data should be reshaped from 1D to 2D matrix due to the usage of convolutional layers. Each row corresponds to an EEG/ECoG sample. At that point, the even rows only should be flipped while keeping the odd rows to exploit both temporal and spatial correlations of the EEG; this approach is called Zigzag, as described in Ref. [16].

Normalization is the next step after the reshaping, as it guarantees a stable convergence of weights and biases while training a machine or deep learning model.

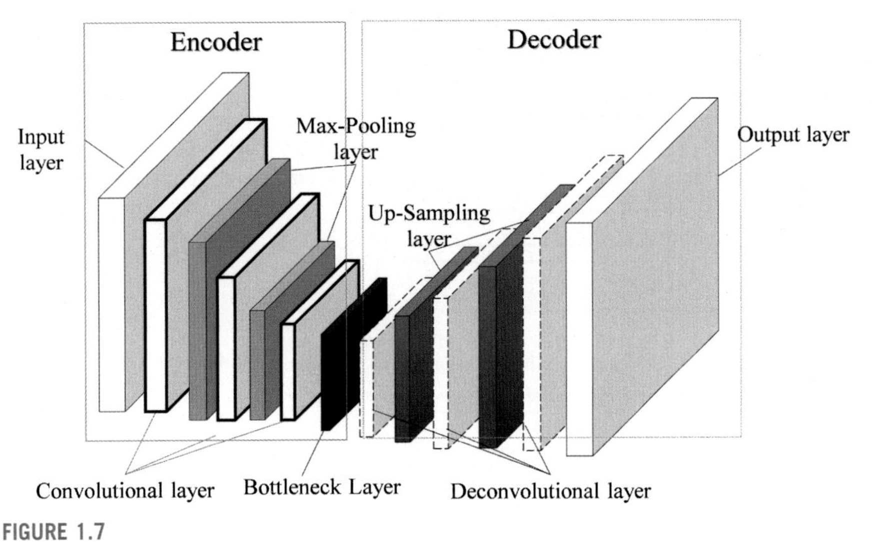

FIGURE 1.7
Representation of convolutional autoencoder.

Besides, unnormalized data may slow and complicate the training step causing internal covariate shift, Eq. (1.1) represents one way of normalizing the data [16].

$$\text{Value after normalization} = \frac{\text{Value before normalization} - \text{minimum value}}{\text{maximum value} - \text{minimum value}} \quad 1.1$$

Similar to any deep neural network, the proposed model contains an input layer, hidden layers, and an output layer. In CAE, the expected output from the last layer should be the same as the input layer. The feature map of each layer, starting from the input layer until reaching the output layer is the input for the next layer. Moreover, the number of filters in each layer changes until reaching to the bottleneck layer, where the number of filters represents the amount of applied compression ratio. Convolutional and max-pooling layers are used in the encoder part of the model, while convolutional and upsampling layers are used in the decoder part. Fig. 1.8 represents the general network architecture used for a particular dataset.

Here is an abstract explanation of each layer in the proposed model. Fully connected layers were not used since the aim of the model is to compress and reconstruct the data rather than classify them.

- **Convolutional 2D layer**: The convolutional 2D layer used in the input, output, and hidden layers.
- **Max-pooling layer**: The max-pooling layers apply a max filter to get the maximum value of a specific region of the input, and then aggregate the maximum values of all the regions into a matrix. Max-pooling layer affects only the width and the height of the input, but the depth stays the same. For example, if the dimension of the input is (x, y) after applying max pooling with nonoverlapping kernel size (k, k), then the dimensions of the output will be $\left(\frac{x}{k}, \frac{y}{k}\right)$

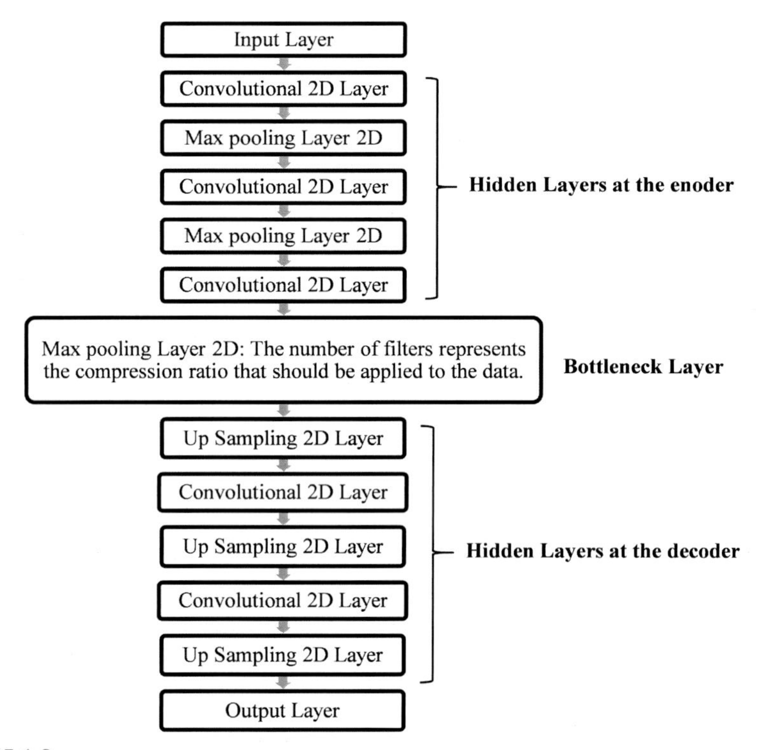

FIGURE 1.8

Deeper representation of the proposed compression approach on certain dataset.

- **Upsampling layer**: The upsampling layer repeats the columns and rows of the data to approximate and reconstruct the original input.

The training of the network is done offline, and once an excellent performance is achieved, the model configuration (matrix) will be saved and used on the edge device of the client.

The calculation of the number of filters at the bottleneck layer is based on the compression ratio. Eq. (1.2) represents the number of filters based on the requested compression ratio, where F should be an integer value. Usually, floor and ceil functions can be used to convert a decimal number to an integer. This conversion may affect the compression ratio, so it should be recalculated based on the decided number of filters using Eq. (1.3). In Eqs. (1.2) and (1.3), F is the number of filters, S is the original size of the data, and $M_i \times N_i$ is the dimensions of the input at the bottleneck layer. The size of the compressed data is represented by the multiplication of the number of filters at the bottleneck layer and the input dimensions to that layer.

$$\text{Number of filters } (F) = \frac{S - (\text{CR} \times S)}{M_i \times N_i} \quad 1.2$$

$$\text{Compression ratio (CR)} = \left|\frac{(F \times M_i \times N_i) - S}{S}\right| \times 100\% \qquad 1.3$$

The size of the dataset and the size of the input in the bottleneck layer decides the maximum compression ratio applied to the signal.

Using CAE as a technique to compress the medical data is a new approach by Ref. [16], which was not addressed before. Different from any other compression technique, deep learning is dataset and application-specific. The same model may not work efficiently for different datasets.

4.2 The optimization of resources in a wireless network

The importance of implementing an adaptive compression based on the network state was essential to manage the dynamics of the wireless network. As mentioned before, the main focus is on minimizing the distortion of the reconstructed data and the transmission energy.

The following parameters were used to solve the optimization problem:

Objective variables

- Distortion $(D)(D)$: The distortion rate should be less than the distortion threshold (D_{th}) (e.g., between 8% and 12%).
- Transmission energy (E_T): Transmission energy per patient at a specific time slot, considering the wireless channel characteristics.

Optimization variables

- Compression ratio (CR)(CR): The used compression ratio at the edge device.
- Data rate (R) (R)

Constraints

- The delay deadline for the end to end delay $(d_{\max})$
- Bandwidth (w): The overall bandwidth should not exceed the maximum bandwidth of the network $(W_{\max})_{max}$.
- Mobility of the patients: the distance between the patient and the base station.

Rayleigh channel model was used to denote the wireless communication environment where both multipath fading and path loss affect the signal to the receiver. The formulation of the multiobjective optimization problem, as in Ref. [17], is explained later.

Eq. (1.4) represents the transmission energy for each patient at a certain time slot while moving. E_T is the transmission energy per patient i, l is the packet length, k_i is the transmitted data, x_i is the channel gain, and w_i and r_i are the bandwidth and the data rate, respectively.

$$E_T = \frac{x_i \times l \times w_i \times k_i \times 2^{\frac{r_i}{w_i}}}{r_i} \qquad 1.4$$

Using regression analysis, the association between the transmitted data and the distortion rate using the CAE model and DWT compression techniques can be represented as in Eqs. (1.5) and (1.6). In Eq. (1.5), D_{CAE} is the distortion rate, c_1 and c_2 are parameters that change based on the dataset. In Eq. (1.6), D_{DWT} is the distortion rate, c_{11}, c_{22}, c_{33}, c_{44}, c_{55}, and c_{66} are statistical model parameters, k_i is the transmitted data, and F is the wavelet filter length [17].

$$D_{CAE} = c_1 k_i^{c_2} \tag{1.5}$$

$$D_{DWT} = c_{11} \cdot F^{-c22} + c_{33} \cdot e^{c_{44} k_i} + c_{55} \cdot (k_i \times 100)^{-c_{66}} \tag{1.6}$$

Below is the multiobjective optimization framework, which manages the tradeoff between the distortion of the reconstructed signal and transmission energy. The objective function normalized by dividing the total value by the maximum value of both the transmission energy and distortion rate, where the maximum value for the distortion rate is 100. The maximum transmission energy ($E_{\max}$) represented in Eq. (1.7). Managing the amount of significance placed for the transmission energy and distortion using λ weighting factor. The value of λ will vary between 0 and 1; when it equals 1, the transmission energy will be considered, and the distortion will be neglected and vice versa. The number of patients (users) is n, $W_{\max}$ is the maximum bandwidth. Eq. (1.8) represents the maximum data rate R_{max} where l is the packet length and d_{max} is the delay deadline [17].

$$E_{max} = \frac{x \times W_{max} \times l \times 2^{\frac{R_{max}}{W_{max}}}}{W_{max}} \tag{1.7}$$

$$\begin{aligned}
&P: min_{CR,R}\left(\sum_{i=1}^{n} \left(\left(\frac{\lambda}{E_{max}} \times E_T \right) + \left(\frac{1-\lambda}{100} \times D \right) \right) \right) \\
&\text{Subject to :} \\
&\quad r_i > 0 \\
&\quad 1 \geq k_i > 0 \\
&\quad w_i > 0 \\
&\quad \sum_{i=1}^{n} w_i < W_{max} \\
&\quad \sum_{i=1}^{n} d_i \leq d_{max} \\
&\quad D_i < D_{th} \\
&\quad R_{max} = \frac{l}{d_{deadline}}
\end{aligned} \tag{1.8}$$

4.2.1 Experiments assessment

The CAE model explained earlier was used for EEG vital signs on a certain dataset; however, when using different datasets, the model will change. We have tested the CAE model on EEG data and ECoG data, where the model outperforms the state-of-art compression/reconstruction approaches.

4.2.2 Datasets

We conduct our experimental analysis on three datasets: two of them are EEG signals, and one is ECoG signals.

4.2.2.1 From the BCI competition IV

- BCI-IV-2a (A) [18]: EEG signals are recorded for different motor imagery tasks using 22 electrodes from nine patients: the imagination of movement of the right and left hands, both tongue and feet. The training size of the data was 6416 samples and 1132 for testing, where the total number of samples is 7548.
- BCI-IV-2b (B) [19]: The EEG signals are recorded from nine patients, in three bipolar recordings. The training size of the data was 5202 samples and 690 for testing, where the total number of samples is 5892.

4.2.2.2 ECoG dataset (C) [20]

- The ECoG signals are recorded to localize the start of the abnormality of brain activity and how it spreads for four epilepsy patients at Harborview Hospital. The training size of the data was 650,200 samples, 260,080 for validating, and 390,120 for testing, where the total number of samples is 1,300,400.

4.2.3 Environment setup

4.2.3.1 Adaptive compression

The CAE model was built in Python programming language as it has professional easy-to-use built-in libraries such as SciPy for scientific computation and scikit-learn, which is a professional grade machine library. **TensorFlow** and **Theano** are two main numerical libraries for building a deep neural network, in Python. We have used TensorFlow, which was released by Google and used directly to build the model or using wrapper libraries built on top of TensorFlow to simplify the process.

Python has a simple library called **Keras**, which runs on top of TensorFlow to facilitate the process of building deep learning for research and development and hide the complexity of TensorFlow. Keras can run on both GPUs and CPU, which make it more powerful and effective because using GPU will reduce the time required for training a model.

As explained earlier, three types of layers were used to build the CAE model, but the number of layers, the hyperparameters, metrics, and loss function change based on the applications.

The datasets from *BCI competition* (*A* and *B*) used the Relu activation function in the hidden layers, and the number of filters varies based on the data size, as explained earlier. The model was trained for 70 epochs with a batch size of 5 [16].

The hidden layers in the *ECoG dataset* (C) used Elu activations function, and the model was trained for 50 epochs with a batch size of 500 because the size of the data was huge.

All models use Adam optimizer, mean square error loss function and mean absolute error, accuracy, and percent-root mean square distortion (PRD) as compilation metrics.

4.2.3.2 Optimization of the network resources

The multiobjective optimization problem was built in MATLAB software using the CVX modeling system for convex optimization. The topology of the tested network consists of 4, 6, and 10 patients following the STEPS mobility model in movement in 4 zones with a minimum speed of 2 m/s and a maximum speed of 6 m/s. The simulation was done using one CAE model as an example that shows the efficiency of integrating the CAE model as an adaptive compression approach in reducing the transmission energy of massive medical data. Table 1.1 represents the used simulation parameters in the multiobjective optimization problem [17].

4.2.4 Results and discussion

4.2.4.1 Adaptive compression

The performance metrics are the following:

- Compression ratio (CR): The size of the dataset decides the maximum compression ratio applied to the data. For example, B and C datasets can reach up to 93% compression, which means only 7% of the data will be transmitted; whereas, in A, the compression rate reaches 98%.
- Percent-root mean square distortion (PRD): S is the size of the data, O_s and R_s correspond to the original signal and the reconstructed signal, respectively.

Table 1.1 The optimization problem simulation parameters.

Parameter	Value	Parameter	Value
Maximum data rate (R_{max})	2000000 bit/s	Packet length (L)	50000 bits
Maximum bandwidth (W_{max})	1.5×10^6Hz	Minimum bandwidth (W_{min})	0 Hz
Noise spectral density(N_0)	-3.98×10^{-21}dBm/Hz	Doppler frequency	0.1 Hz
Time steps	60 s	Distortion threshold	Between 8% and 16%
Delay deadline	0.025 s	F (filter length)	2
c_1	4.9404	c_2	−0.351
c_{11}	2.2	c_{44}	1
c_{22}	0.3	c_{55}	3620
c_{33}	1.475	c_{66}	1.465

$$\text{PRD}\ (\%) = \sqrt{\frac{\sum_{i=1}^{S}(O_s - R_s)^2}{\sum_{i=1}^{S} O_s^2}} \times 100\% \qquad 1.9$$

- Mean absolute error (MAE):

$$\text{MAE}(\%) = \left| \frac{\sum_{i=1}^{S} O_s - R_s}{S} \right| \times 100\% \qquad 1.10$$

4.2.4.1.1 ***Compression and distortion relation*** Fig. 1.9 represents the relation between the distortion rate and the compression ratio. The usage of CAE architecture facilitates applying high compression ratios while keeping the distortion as minimum as possible. In A dataset, at 98% compression, the PRD reached 1.3% only owing to the data itself. However, in B dataset, when sending 7% of the data, the PRD reaches almost 12%, which is extremely good compared with other techniques. For dataset C, at 94% compression ratio, the PRD reached 11.7%.

4.2.4.2 Optimization of the network resources

The duration of the simulation equals 60 time slots. Each patient has some sensor nodes that collect the data and send them to the PDA. The channel variations were modeled using Flat Rayleigh fading with a Doppler frequency of 0.1 Hz. The location of the patient relative to the base station will frequently change during the simulation time 60 time slots. The data length and the delay deadline are the same for all the patients.

4.2.4.2.1 ***Weighting factors effect*** The weighting factor used in these runs will vary between 0 and 1, where 1 means that transmission energy is only considered while 0 means that full privilege is given to distortion. Fig. 1.10 shows the effect of changing the weighting factors on the average distortion rate and average

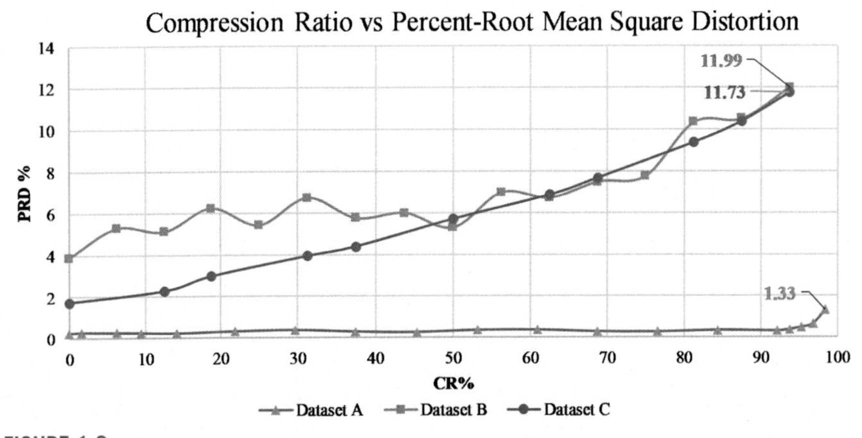

FIGURE 1.9

Relation between CR (%) and PRD (%).

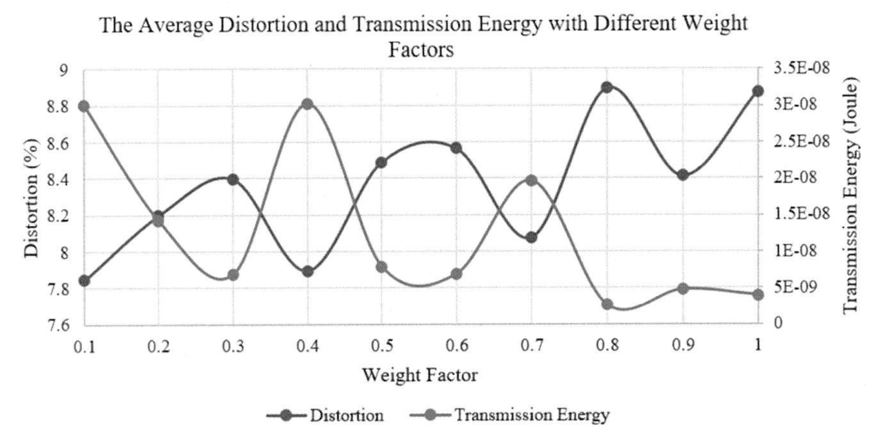

FIGURE 1.10

Relation between the average distortion and average transmission energy at different weighting factors.

transmission energy for 4, 6, and 10 patients. The distortion relation with respect to the transmitted data from B dataset used in the optimization problem. In the simulation, five different distortion thresholds were used, starting from 8% up to 16% by incrementing the threshold by 2% at each iteration. The optimization problem runs for 300 times, and it consumed different time based on the number of patients because running the optimization problem for four users takes around 10 minutes running while having 10 users takes around 80 minutes.

Fig. 1.11 illustrates the relationship between weighting factors and the transmission energy when using CAE as a compression approach and DWT. CAE ensures minimizing the transmission energy much more than DWT, while it reconstructs the medical data that will have distortion less than any of the state-of-the-art techniques. Fig. 1.12 expresses the power of CAE compared with DWT on both the distortion and transmission energy, as for both of them CAE was able to outperform DWT [17].

4.2.4.2.2 ***Channel bandwidth effect*** Considering the bandwidth as a decision parameter has a significant effect on solving the optimization problem. The channel allocated for each patient specifies the amount of bandwidth for its patient, where the overall bandwidth consumed by all patients should not go beyond the bandwidth threshold. Considering the bandwidth for all the patients have a significant effect on the distortion of the reconstructed signal and the transmission energy. Fig. 1.13 demonstrates the average distortion and the average transmission energy for three scenarios, where the number of patients equals 4, 6, and 10 patients. However, for distortion, when the number of patients is few, the bandwidth will have a definite effect, but it is not the case when the number of patients increases.

On the other hand, considering the bandwidth gives an exceptional implication to the system concerning transmission energy compared with having constant

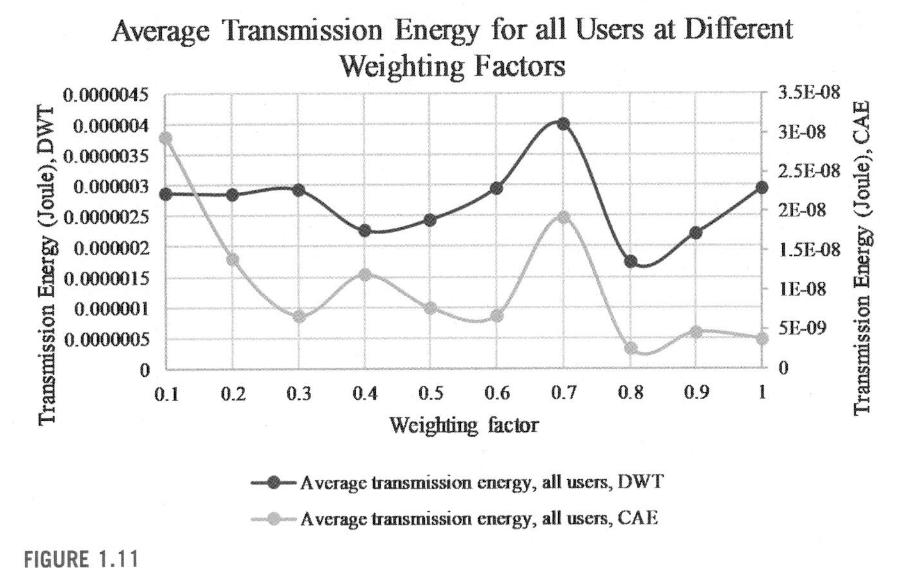

FIGURE 1.11

Average transmission energy when using DWT and CAE as an adaptive compression technique for all patients (users).

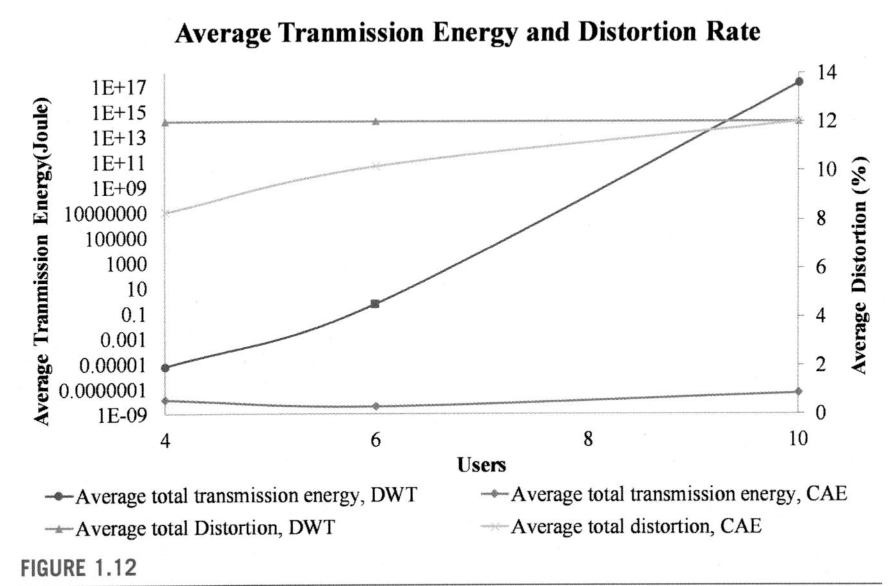

FIGURE 1.12

Average transmission energy and distortion rate when using DWT and CAE as an adaptive compression technique for all patients (users).

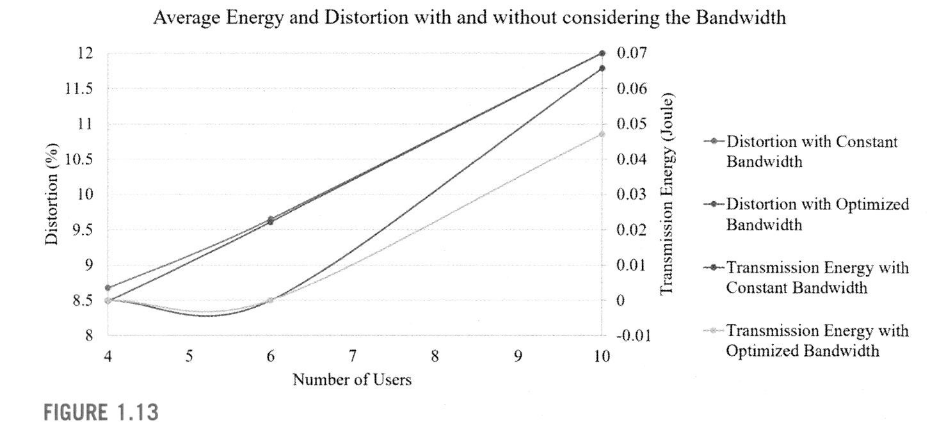

FIGURE 1.13

The average transmission energy and the average distortion when the number of patients (users) equals 4, 6, and 10.

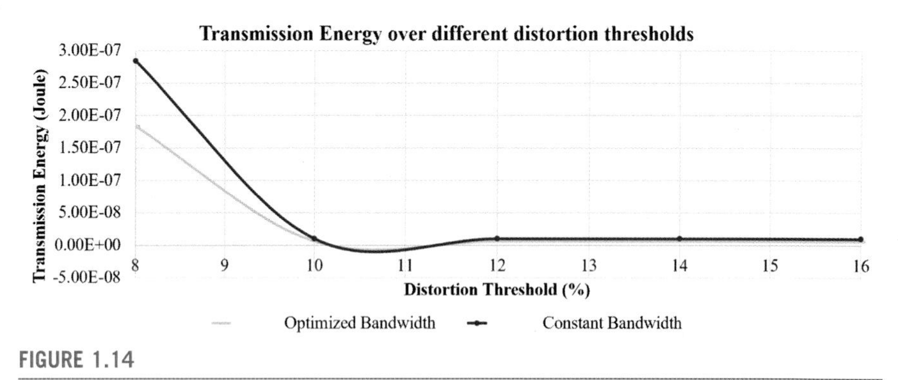

FIGURE 1.14

The average transmission energy for a particular patient out of six patients through five different distortion thresholds.

bandwidth for all the patients. Fig. 1.14 shows the transmission energy at different distortion thresholds for one out of six current patients involved in the m-Health system. A noticeable difference between transmission energy when considering bandwidth as a decision parameter in the optimization problem and when setting the bandwidth to a constant value, which results in increasing the productivity of the system by almost 54.3% [17].

4.2.4.2.3 ***Distortion threshold effect*** The distortion threshold limits the amount of compression applied to the medical data; owing to that increasing the distortion threshold accommodates the compression ratio, as it will increase, causing a decline in the transmission energy. Fig. 1.15 illustrates the association between the distortion threshold and transmission energy. For example, at 16% distortion threshold, the transmission energy decreases sharply because the data were compressed with a ratio higher than when having a distortion threshold equal to 8%. Fig. 1.15 ensures that the distortion and transmission energy are two conflicting objectives of the optimization problem [17].

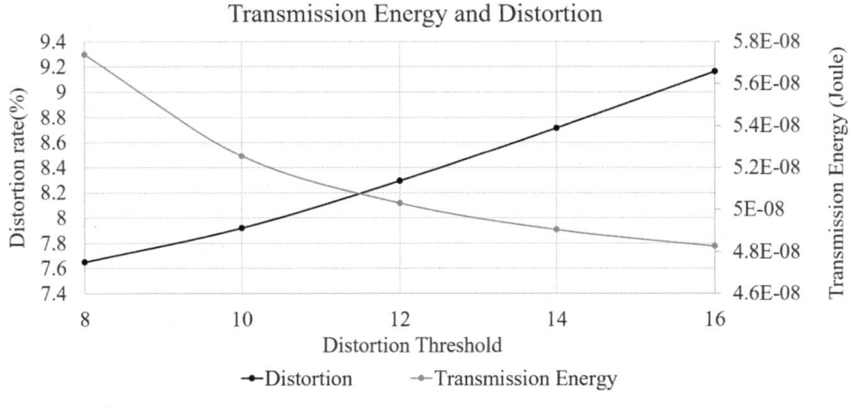

FIGURE 1.15

The transmission energy and distortion rate at different distortion thresholds.

5. Summary

Recently, the amount of transmitted medical data to the caregiver in the m-Health system increases swiftly. Therefore, the need for an optimal resource allocation framework with respect to both the application and network requirements is an essential target. Deep learning is an effective way to manage the preprocessing of the data before being sent through the network, and it can outperform the state-of-the-art preprocessing techniques where compression is one example of data reduction preprocessing approach.

The primary outcomes of this chapter can be summarized as follows:

- Deep learning is different from all other signal processing and compression techniques, as it applies compression based on the previous learning phase.
- The more training applied to the neural network with more data, the less the distortion, and this is the known rule in deep learning, which proved through the experimental phase of building the model.
- Deep learning is application-specific. The same model cannot be used for different datasets, even if they contain the same data type. This point was proved by using the model of A dataset with C dataset, where we end up with horrible results, especially in visualization.
- Using deep learning may lead to minimizing the distortion while maximizing the compression ratio for some applications. For example, in the case of A dataset, the model configurations can be used in the network without adaptive compression and regardless of the network state, where we apply the maximum compression ratio at any time. However, for other datasets like B, good results but not optimized were obtained. Therefore, optimization with respect to different network states should be applied to adjust the compression ratio considering the network state and the application requirements.

- Considering the bandwidth as a decision parameter while solving the optimization problem increases the efficiency of the system by more than 50%. Both the distortion and the transmission will be minimized effectively compared with a scenario that gives the same bandwidth to all users despite their network state, such as the allocated channel state.
- The worse the users' channel, the more allocated bandwidth, and then more transmission energy will be consumed.
- Increasing the compression ratio means increasing the distortion of the reconstructed signal, which implies a decrease in the transmission energy as well as the allocated bandwidth.
- Changing the weighting factor to manage the trade-off between distortion of the reconstructed signal and transmission energy gives a definite conclusion that when giving the highest priority to the transmission energy, the distortion will reach the maximum possible value and vice versa.

Acknowledgments

This work was made possible by Qatar University Grant QUHI-CENG-19/20-1. The work of Abeer Al-Marridi has been supported by GSRA grant # GSRA5-1-0326-18026 from the Qatar National Research Fund (a member of Qatar Foundation). The findings achieved herein are solely the responsibility of the authors.

References

[1] Global diffusion of eHealth: making universal health coverage achievable. Report of the third global survey on eHealth. Geneva: World Health Organization; 2016. http://apps.who.int/iris/bitstream/hand le/10665/252529/9789241511780-eng._pdf.

[2] mHealth: new horizons for health through mobile technologiesGlobal Observatory for eHealth series, vol. 3. Geneva: World Health Organization; 2011. http://www.who.int/goe/publications/goe_mhealth_web.pdf.

[3] mHealth: use of appropriate digital technologies for public health: report by the director-general. Geneva: World Health Organization; 2018. A71/20, http://apps.who.int/gb/ebwha/pdf_files/WHA71/A71_20-en.pdf.

[4] Lee J. Smart health: concepts and status of ubiquitous health with smartphone. In: ICTC 2011. IEEE; September 2011. p. 388–9.

[5] A wearable sensor-based activity prediction system to facilitate edge computing in smart healthcare system.

[6] Li K, Shao MW, Wu WZ. A data reduction method in formal fuzzy contexts. International Journal of Machine Learning and Cybernetics 2017;8(4):1145–55.

[7] Quick D, Choo KKR. Big forensic data reduction: digital forensic images and electronic evidence. Cluster Computing 2016;19(2):723–40.

[8] Wang J, Yue S, Yu X, Wang Y. An efficient data reduction method and its application to cluster analysis. Neurocomputing 2017;238:234–44.

[9] ur Rehman MH, Liew CS, Abbas A, Jayaraman PP, Wah TY, Khan SU. Big data reduction methods: a survey. Data Science and Engineering 2016;1(4):265–84.
[10] Awad A, Amal S, Ali J, Mohamed A, Chiasserini C-F. In-network data reduction approach based on smart sensing. In: Global communications conference (GLOBECOM), 2016. IEEE; December.2016. p. 1–7.
[11] Ravì D, Wong C, Deligianni F, Berthelot M, Andreu-Perez J, Lo B, Yang GZ. Deep learning for health informatics. IEEE Journal of Biomedical and Health Informatics 2017;21(1):4–21.
[12] Gibson A., Patterson J. Deep Learning; n.d. Available from: https://www.safaribooksonline.com/library/view/deep-learning/9781491924570/ch04.html.
[13] Hussein R, Mohamed A, Alghoniemy M. Scalable real-time energy-efficient EEG compression scheme for a wireless body area sensor network. Biomedical Signal Processing and Control 2015;19:122–9.
[14] Brownlee J. Deep learning with Python: develop deep learning models on Theano and TensorFlow using Keras. Place of publication not identified: Machine Learning Mastery; 2016.
[15] Said AB, Mohamed A, Elfouly T. Deep learning approach for EEG compression in mHealth system. In: 2017 13th international wireless communications and mobile computing conference. IWCMC; 2017.
[16] Al-Marridi AZ, Mohamed A, Erbad A. Convolutional autoencoder approach for EEG compression and reconstruction in m-health systems. In: 2018 14th international wireless communications & mobile computing conference (IWCMC). IEEE; June 2018. p. 370–5.
[17] Al-Marridi AZ, Mohamed A, Erbad A, Al-Ali A, Guizani M. Efficient EEG mobile edge computing and optimal resource allocation for smart health applications. In: 2019 15th international wireless communications & mobile computing conference (IWCMC). IEEE; June 2019. p. 1261–6.
[18] Brunner C, Leeb R, Müller-Putz G, Schlögl A, Pfurtscheller G. BCI Competition 2008–Graz data set A. Institute for Knowledge Discovery (Laboratory of Brain-Computer Interfaces), Graz University of Technology; 2008. p. 136–42.
[19] Leeb R, Brunner C, Müller-Putz G, Schlögl A, Pfurtscheller G. BCI Competition 2008–Graz data set B. Austria: Graz University of Technology; 2008.
[20] Miller KJ. A library of human electrocorticographic data and analyses. Stanford Digital Repository; 2016. Available from: https://purl.stanford.edu/zk881ps0522.

CHAPTER

2 Applying an efficient evolutionary algorithm for EEG signal feature selection and classification in decision-based systems

Sajjad Afrakhteh, Mohammad Reza Mosavi

School of Electrical Engineering, Iran University of Science and Technology, Narmak, Tehran, Iran

1. Introduction

The expansion of medical knowledge and the complexity of diagnosis and treatment decisions have drawn the attention of experts to the use of medical decision support systems. The use of different types of smart systems in medicine is increasing. So, today the impact of different types of smart medical systems has been studied. Intelligent systems (expert and neural network) have structures, components, and capabilities that improve decision-making. Artificial intelligence helps the physician to consider more and more variables when diagnosing the disease or choosing treatment. Despite the many advantages, the use of artificial intelligence systems in medicine faces many serious obstacles and challenges. Among these limitations, we can indicate the complexity and accuracy challenges. Therefore, we used two applications including epilepsy diagnosis (for feature selection problem) and motor imagery classification (for classification problem). The electroencephalography (EEG) datasets are used for evaluation. In these areas, scientists have done many kinds of research works to enhance the performance of decision-based systems. Here, we review some of the works. Epilepsy is considered one of the most important neurological diseases in the world, which can affect all age groups. Epilepsy and neurological attacks affect nearly 2.5 million Americans of all ages [1]. Overall, 1%–3% of the US population is affected. Sudden deaths occur in about 19 per 1000 of epileptic patients in a year [1]. Despite the methods and options available to treat epilepsy, about one-third of patients still have indomitable seizures. Patients with indomitable epilepsy are at greater risk of injury and death. Also, patients with epilepsy suffer from social problems and low levels of quality of life in the community, as well as marriages, higher education, and a higher standard of living. With these descriptions, it is always important to recognize epilepsy; therefore, many researchers have been looking for a method to better diagnose this phenomenon that interdisciplinary researchers apply their algorithms for improving the performance

Energy Efficiency of Medical Devices and Healthcare Applications. https://doi.org/10.1016/B978-0-12-819045-6.00002-9

of the detection system. These algorithms include machine learning and signal processing. In Ref. [2], Srinivasan et al. used artificial neural networks (ANNs) to diagnose epilepsy using EEG signals. They used two types of Elman neural networks (NNs) and probability-based NNs. Ocak in Ref. [3] presented a technique based on wavelet transform and entropy analysis of EEG signals for seizure detection. In their study, the signals of the approximate coefficients were analyzed using discrete wavelet transform (DWT). Then, the entropy of each of these coefficients was determined, and based on these values, they were able to detect epileptic seizures with accuracy above 96%. Tzallas et al. [4], with the argumentation that the evolution of an epileptic seizure is often dynamic and nonstatic, the EEG signals consist of different frequencies, and given the limitations of frequency-based methods, presented an analytical method based on time-frequency information for the epileptic seizure classification. Acharya et al. [5] categorized a method for automatic detection of normal, preictal, and ictal conditions using different classifiers, and their results showed that a fuzzy classifier was able to separate three classes with the accuracy of 98.1%. Subasi classified epilepsy using wavelet features and a modular NN structure [6]. This network structure achieved accuracy rates that were higher than that of the stand-alone NN model. Acharya et al. developed a classification method for the diagnosis of epilepsy based on deep learning, using convolutional neural networks (CNNs) to analyze EEG signals. Their proposed method resulted in an accuracy of 88.67%. In Ref. [7], Subasi et al. first decomposed the EEG signals into different frequency bands, then lowered the dimensionality using data reduction techniques such as principal component analysis, independent component analysis, and linear discriminant analysis (LDA). Finally, these features were applied to the input of a support vector machine (SVM) classifier. The results of the classification indicate the good performance of their proposed method. Martis et al. presented the empirical mode decomposition for analyzing EEG signals to diagnose epilepsy [8]. In Ref. [9], a method based on K-means clustering and multilayer perceptron NNs (MLP-NN) was presented for seizure detection. Sharma et al. classified epilepsy using phase and space combination of intrinsic mode functions of EEG signals [10]. Shivnarayan et al. proposed a new technique based on features extracted using tunable wavelet transform and Kraskov entropy [11]. In Ref. [12], Kai et al. proposed a technique based on the Hilbert marginal spectrum for seizure detection. A method based on entropy-based features was proposed in Ref. [13]. The discriminative property of their proposed method is the low complexity that makes it suitable for real-time applications. Another application is studied in this paper, which is one of the main categories of brain–computer interface (BCI) systems, is the motor imagery classification. It is a very important and favorite application for people with spinal cord disabilities who are unable to move [14,15]. These systems help people with disabilities to better communicate with their surroundings and gain control of their lives in conveying their perceptions to others. Further development of the BCI systems depends on the highly accurate and low complex design. One of the newest concepts in machine learning is the combination of NNs and evolutionary algorithms (EAs). This concept allows the best parameter setting and accuracy to be achieved. In Ref. [16], we used the PSOGSA algorithm to train MLP-NNs parameters. In addition, we used the new version of the particle swarm optimization (PSO) algorithm to train ANFIS parameters to achieve higher performance [17]. In this paper, a new version of PSO is introduced and is used for feature selection and MLP-NN training. Before applying it on the diagnosis problem, it is used for finding the global optimum of 14 standard benchmark functions, and its results compared to other EAs, such as PSO [18], simulated annealing (SA) [19], differential evolution (DE) [20], and genetic algorithm (GA) [21], in terms of finding the global minimums. After demonstrating the superiority of the proposed optimization algorithm in finding the minimum value of unimodal and multimodal benchmark functions, it is used for feature selection, after feature extraction phase, and MLP-NN training, in the classification phase, to gain higher performance of the diagnosis system. In the classification phase, instead of using the backpropagation algorithm to update weights and thresholds, the proposed Mo-PSO is used to avoid trapping in local minima.

The remainder of the paper is as follows. Section 2 describes the datasets used in this paper. In Section 3, the proposed method is presented and the process of applying the method to select epileptic features and classify motor imagery is discussed. The results of the suggested approach on benchmark functions and the two applications are presented and discussed in Section 4. Finally, in Section 5 we present the conclusions of the paper.

2. Datasets

This section describes the data used in this article. Here are the epilepsy and the motor imagery datasets to consider.

2.1 Epilepsy

Appropriate datasets for the data used in this study are part of the data for two different categories of patients' normal and seizure state. The data were provided by the Epilepsy Department of the University of Bonn Germany and are available online [22]. Each group has 100 single-channel EEG signals, with 2.36 s length, and sampled at the frequency of 107.61 Hz. Each part of the data is considered as a separate EEG. The first stage is the filtering that is done by a third Butterworth filter [23], which aims to analyze the EEG signals with a frequency range of 4−32 Hz, including theta (4−8 Hz), alpha (8−13 Hz), and beta (14−30 Hz) waves. The EEG signals for seizure and health state of an epilepsy patient, before and after filtering, are shown in Fig. 2.1. After the filtering process, we divide the data for each class into 205 different trials and assign a label to each trial, indicating the class corresponding to each of these trials.

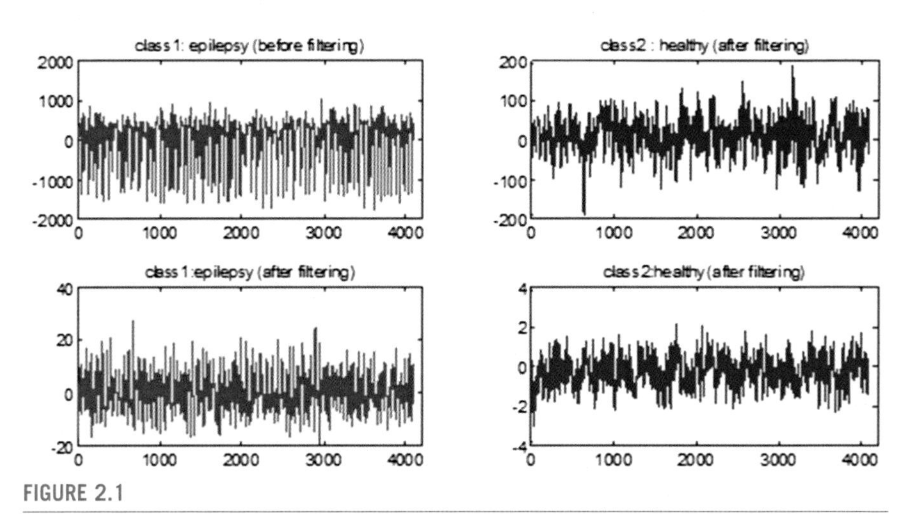

FIGURE 2.1

EEG signal for seizure and health state of an epilepsy patient, before and after filtering.

2.2 Motor imagery classification

In addition, to examine the motor imagery classification, the BCI Competition IV calibration dataset, which is a two-class dataset, is used [24]. The data were recorded using the appropriate sensors from 59 different positions, which correspond to seven different subjects and represent the left hand and right foot motor imagery. For motor imagery classification, the desired frequency band is 8–30 Hz. This frequency band contains alpha (8–13 Hz) and beta (14–30 Hz) waves. These frequency bands are shown in Fig. 2.2. So, a band-pass filter is used for the filtering process that passes

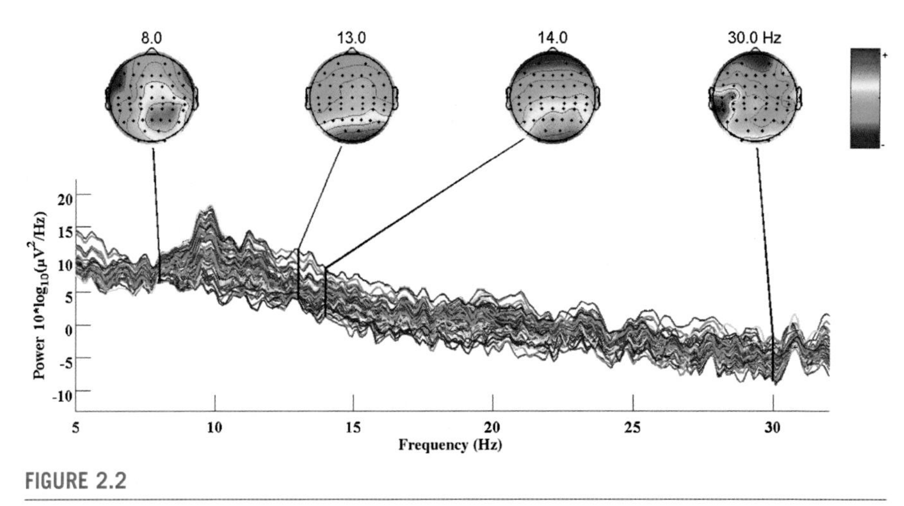

FIGURE 2.2

Frequency band of motor imagery dataset.

the desired frequency band and rejects others. The type of the filter is third Butterworth.

3. Methods

In this section, we explain the techniques employed, namely the use of the proposed EA in each of the two applications (feature selection in epilepsy and classification of motion imagery), and then we present the proposed EA.

3.1 Feature selection in epilepsy diagnosis

In addition to the accuracy in a diagnostic system, the complexity is another crucial parameter that is of interest to designers of diagnostic systems. Therefore, a system with low complexity and high accuracy is always desirable, and there is often a tradeoff between these two criteria. As stated in the previous section, each time the data are recorded, the signals are recorded from 100 different electrodes. The purpose of this paper is to obtain a highly accurate system with fewer signals, which is very valuable for diagnostic systems. By this idea, the distribution of electrodes on the patient's head is changed and power consumption can be reduced significantly. The proposed method for this signal selection is a novel modified PSO algorithm and is called Mo-PSO. This algorithm is presented in the following section and is applied on sum unimodal and multimodal standard benchmark function to evaluate its performance in terms of finding the global optima. By applying the proposed algorithm on the filtered signals and choosing a proper fitness function, the optimal signals or features are selected and the rest of the signals, that do not contain useful information and only increase the system complexity, are eliminated. Finally, the selected signals are the input of the NNs in the classification phase to detect the normal or seizure state of the subject. The block diagram of the proposed structure is presented in Fig. 2.3. As can be seen, by using the optimal features, the classifier determines the output state based on lower feature number and this leads to a low complex diagnosis system.

3.1.1 Fitness function

After that, all solutions are generated based on the Mo-PSO, the fitness of subsets produced by the search agents is evaluated by an NN that is trained and tested for data with the space of features related to these agents. So, the fitness function is defined as Eq. (2.1):

$$Fitness_i = (\alpha \times MSE_i) + \left((1-\alpha) \times \frac{S_i}{N}\right) \tag{2.1}$$

In the above relation, the MSE_i is the classification error corresponding to the reduced data with the feature space related to the factor ith and S_i is the number of features in the subset of the feature selected by the ith agent of the proposed

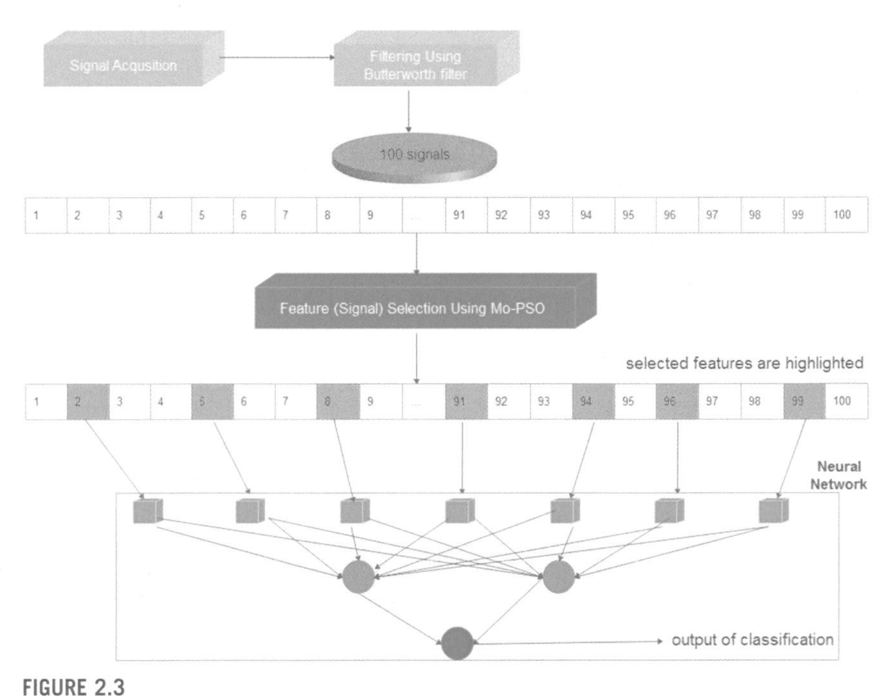

FIGURE 2.3

Flowchart of the proposed technique for an epilepsy diagnosis.

Mo-PSO algorithm. As the main purpose of minimizing the classification error is in the feature selection problem, the α value is considered to be close to one.

3.2 Proposed method for motor imagery classification

First, the filtered signals are applied to the common spatial pattern (CSP), and then the extracted features are the inputs of the NN classifier. So, for better understanding, these two parts including the CSP feature extraction and the NN are described here.

3.2.1 CSP feature extraction

This strategy of CSP was first introduced by Ramsor to classify motor imagery [25]. The fundamental thought of the method is to employ a linear relationship to transfer a multichannel EEG dataset to a subspace with a lower dimension by a matrix that each row of which is made up of the weights of each channel. This transformation can enhance the difference between the two classes. The CSP technique uses the concept of the diagonalization of the covariance matrix for both groups simultaneously. Let us explain the details of this algorithm. We assume that the data matrices of the EEG signal are related to the first class (left-hand motor imagery) and the second class (right foot motor imagery). Dimensions of this matrix are

$N * T$, in which N is the number of channels and T is the number of samples in each channel. The normalized covariance matrix for each class is given in Eq. (2.2):

$$C_1 = \frac{X_1.X_1^T}{trace\left(X_1.X_1^T\right)} \quad , \quad C_2 = \frac{X_2.X_2^T}{trace\left(X_2.X_2^T\right)} \tag{2.2}$$

where X^T is the transpose of the matrix X, and trace (.) is defined to be the sum of the elements on the main diagonal of matrix A. The average normalized covariance of $\overline{C}_1$, $\overline{C}_2$ obtained by averaging the normalized covariance in all trials. The combination of these covariance matrices together can be decomposed as Eq. (2.3):

$$C = \overline{C}_1 + \overline{C}_2 = U_0 \sum U_0^T \tag{2.3}$$

where U_O is the eigenvectors and $\sum$ is a diagonal matrix of eigenvalues of the covariance matrix C. Then, a conversion ($P = \sum^{-1/2} U_0^T$) is applied to the average covariance matrices for both classes as Eq. (2.4):

$$S_1 = P\,\overline{C}_1\,P^T \quad , \quad S_2 = P\,\overline{C}_2\,P^T \tag{2.4}$$

S1 and S2 have the same eigenvectors, and the combination of the corresponding eigenvalues for both classes is equal to one. In other words, it can be written as Eq. (2.5):

$$S_1 = U \sum\nolimits_1 U \quad , \quad S_2 = U \sum\nolimits_2 U \quad \text{and} \quad \sum\nolimits_1 + \sum\nolimits_2 = I \tag{2.5}$$

Therefore, the eigenvectors for S_1 have the highest value, the same vectors for S_2 have the lowest value and vice versa. As a result, the matrix of weights is obtained as Eq. (2.6):

$$W = U^T P \tag{2.6}$$

In this algorithm, we arrange the matrix of eigenvalues in descending order. Therefore, we choose the first m weights and the last m weights, which correspond to the maximum and minimum values, and we leave the rest of the weights as zeros. Therefore, the EEG data can be converted into the independent components as Eq. (2.7):

$$Y = W\,X \tag{2.7}$$

The resulting matrix in a new subspace is a $2m * T$ matrix.

3.2.2 Neural networks as a classifier

One of the important applications in diagnosis systems is the motor imagery classification. For this application, the EAs are used in the classification phase to optimize the weights of the NNs. NNs is the supervised machine-learning algorithm. McCulloch and Pitts started the study of NNs in 1943 [26]. Rosenblatt founded single-layer networks (SLNs) in 1962, which were called perceptron [27]. In the 1960s, it was empirically shown that perceptron was capable of solving many simple problems, but they could not solve many complex problems [28]. With the discovery of the backpropagation (BP) algorithm by Rumelhart et al. in 1988, new studies on NNs have resumed [29]. The special importance of this algorithm was that it could train the MLP-NNs. The study of ANNs is largely inspired by natural learning systems in which a complex set of interconnected neurons is contributed to the learning task. Neurons naturally connect in a specific way to form an NN. The connection way of neurons can be in an SLN or MLP-NN. MLP-NN consists of an input layer, an output layer, and one or more layers between them (the hidden layer) that are not directly connected to the input data and output results. The input layer units are simply responsible for distributing the input values to the next layer. They do not affect the input signals. For this reason, the number of layers was not counted. The network consists of an output layer that responds to the input signals. The number of neurons in the input layer and the output layer is equal to the number of inputs and outputs and the hidden layer(s) have the task of connecting the input layer to the output layer. With these hidden layers, the network can extract nonlinear relationships from the data presented to the network. The purpose of BP network training is to strike a balance between learning and generalization capabilities. Learning ability means responding appropriately to input patterns used for training, and generalization means good response to the similar inputs that are not exactly identical. One of the disadvantages of the BP algorithm is trapping in local minima that leads to inefficient parameter settings. For solving this issue, the stochastic-based method is proposed for the best parameter setting. The proposed algorithm for updating the NN's parameters is the Mo-PSO algorithm. After optimizing the NN's parameters, it is used for classification. Fig. 2.4 shows the proposed method in the classification

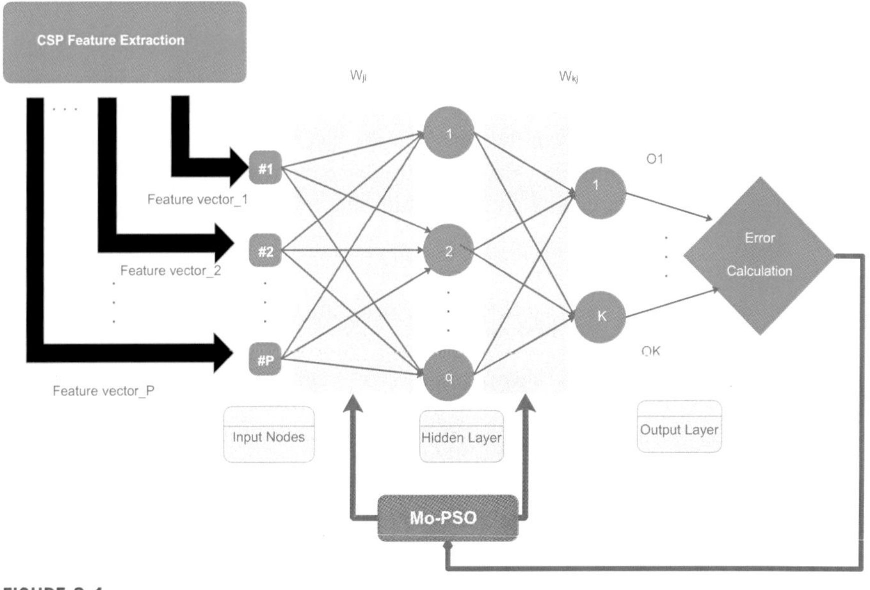

FIGURE 2.4

Block diagram of the proposed method for motor imagery classification.

phase for motor imagery feature vectors obtained by the CSP feature extraction method. As can be seen, the outputs of NNs are compared with the target values. Then, the results of this comparison are expressed as an error. The proposed EA, Mo-PSO, uses this error to optimize the parameter setting. The following is a mathematical expression of this process.

The output of the jth neuron of the hidden layer can be formulated as Eq. (2.8):

$$S_j(n) = \sum_{i=1}^{P} w_{ji}(n)x_i(n) \quad , \quad j = 1, 2, \dots, q \tag{2.8}$$

where $S_j(n)$is the output of the jth neuron in the hidden layer, $w_{ji}(n)$ is the connection weights between input nodes and hidden layer, $x_i(n)$ is the ith input (feature) vector, and n is the time index.

After the hidden-layer computations, the output of neurons in the output layer is calculated as Eq. (2.9):

$$O_k(n) = \sum_{i=1}^{P} w_{kj}(n) \times f(S_j(n)), \quad k = 1, 2, 3, \dots, K \tag{2.9}$$

where $w_{kj}(n)$ is the weights that connect the hidden layer to the output layer, $f(.)$ is the activation function, and K indicates the number of neurons in the output layer. So, the fitness function can be described as Eq. (2.10):

$$error(k) = \sum_{i=1}^{K} (o_i(k) - d_i(k)) \rightarrow fitness = error = \sum_{k=1}^{Train\_num} \frac{error(k)}{Train\_num} \tag{2.10}$$

In the optimization problem, parameters that should be optimized is the weight vector of the NN. So, the parameters can be obtained by minimizing the fitness function:

$$\{w_{ji}, w_{kj}\} = \arg\ \min \quad (fitness) \tag{2.11}$$

In this paper, the Mo-PSO algorithm is purposed to solve the minimization problem and find the best values of these parameters. In the next section, the proposed EA, Mo-PSO, is introduced.

3.3 Mo-PSO algorithms

PSO is one of the most powerful smart optimization algorithms in the field of swarm intelligence. The algorithm, introduced by James Kennedy and Eberhart in 1995, is inspired by the social behavior of animals that live in small and large groups [18]. In this model, simple behaviors, nearest neighbors, and speed adjustment have been implemented. In the PSO algorithm, the particles are randomly placed in a search space, and in each iteration, the nearest neighbor of the particle is selected and the particle velocity is replaced by the velocity of its nearest neighbor. This allows the group to quickly converge in an indeterminate direction without change. In the

PSO algorithm, the population particles interact directly and solve the problem by exchanging information and recalling good memories of the past. The velocity vector and the position vector in the search space define each particle in the group. At each iteration, the new particle position is updated according to the current velocity vector, the best position found by that particle (pbest), and the best position found by the best particle (gbest) in the group. Suppose we have a d-dimensional search space. The position vector X_i describes the ith agent in the search space as Eq. (2.12):

$$X_i = (x_{i1},\ xi_{i2},\ \ldots, xi_{id}\ ) \tag{2.12}$$

The velocity vector of the ith particle is also defined by the vector as Eq. (2.13):

$$V_i = (v_{i1},\ vi_{i2},\ \ldots, vi_{id}) \tag{2.13}$$

The best position that the ith agent finds is defined by $P_{i,best}$, and this shows the local experience:

$$P_{i,best} = (p_{i1},\ p_{i2}, \ldots, p_{id}) \tag{2.14}$$

The best position that is found by the best agent among all particles is gbest, and this indicates the global experience:

$$P_{g,best} = \left(p_{g1}, p_{g2}, \ldots, p_{gd}\right) \tag{2.15}$$

With these parameters, the velocity vector for each particle is calculated as Eq. (2.16):

$$V_i(t) = w * V_i(t-1) + \underbrace{c_1 * rand_1 * (pbest(i) - X_i(t-1))}_{local} \underbrace{+c_2 * rand_2 * (gbest - X_i(t-1))}_{Global} \tag{2.16}$$

where W is the inertial weight coefficient (moving in its own direction), indicating the effect of the previous iteration velocity vector on the velocity vector in the current iteration. c_1 and c_2 are the cognitive factors for local and global experience, respectively. *rand1* and *rand2* are random numbers with uniform distribution in the range of 0−1. In addition, $V_i(t-1)$ is the velocity vector and $X_i(t-1)$ is the position vector, for ith particle, in the previous iteration.

Therefore, each agent updates its position depending on its velocity vector as Eq. (2.17):

$$X_i(t) = X_i(t-1) + V_i(t) \tag{2.17}$$

So far, the general principles of the PSO algorithm were described. In this paper, for obtaining better performance in approaching the global minimum, a new version of the PSO algorithm (Mo-PSO) is introduced. The main idea of the Mo-PSO technique is to add another term to the calculation process of the particle velocity. Our main idea is to compare the local and global experience for velocity updating.

This information is useful to improve convergence speed and reaching the global point. With these interpretations, the velocity is calculated as Eq. (2.18):

$$V_i(t) = w * V_i(t-1) + \underbrace{c_1 * rand_1 * (pbest(i) - X_i(t-1))}_{Local\ Experience} + \underbrace{c_2 * rand_2 * (gbest - X_i(t-1))}_{Global\ Experience} + \underbrace{H(pbest(i), gbest)}_{New\_Term\ that\ combines\ local\ and\ global} \tag{2.18}$$

where this combination has the subtraction form as Eq. (2.19):

$$H(pbest(i), gbest) = c_3 * rand_3() * (gbest - pbest(i)) \tag{2.19}$$

The method proposes a modification in velocity, so compare the global best to the local best. With this change, in each iteration, this comparison is done and leads to a better estimation of the velocity for each particle. Fig. 2.4 shows this process for updating velocity. Finally, the updated velocity is placed in Eq. (2.19) to obtain an updated position.

3.3.1 Implementation steps of the Mo-PSO algorithm

The implementation steps of Mo-PSO algorithm are as follows:

Step 1: Random generation of the initial population of particles.

The random generation of the initial population (RGIP) is simply the random determination of the initial location of the particles by a uniform distribution in the solution space (search space). The RGIP step exists in almost all probabilistic optimization algorithms. However, in this algorithm, in addition to the initial random location of the particles, a certain amount of initial particle velocity is also assigned. The proposed range for the initial particle velocity can be deduced from Eq. (2.20):

$$\frac{\min(X) - \max(X)}{2} \leq V \leq \frac{\max(X) - \min(X)}{2} \tag{2.20}$$

The initial population size is determined depending on the application. In general, the number of primary particles is a compromise between the parameters involved in the problem. Empirically, selecting an initial population size of 20–30 particles is a good choice for almost all test problems. We can also consider the number of particles a little more than we need to have a margin of safety in the face of local minimums (Fig. 2.5).

Step 2: Evaluate the fitness function (cost function) for the particles.

At this stage, we must evaluate each agent, which describes a solution in the problem under investigation.

Step 3: Recording the best location for each agent (Pbest(i)) and the best location among all agents (gbest)

At this point, depending on the iteration number, there are two cases to consider:

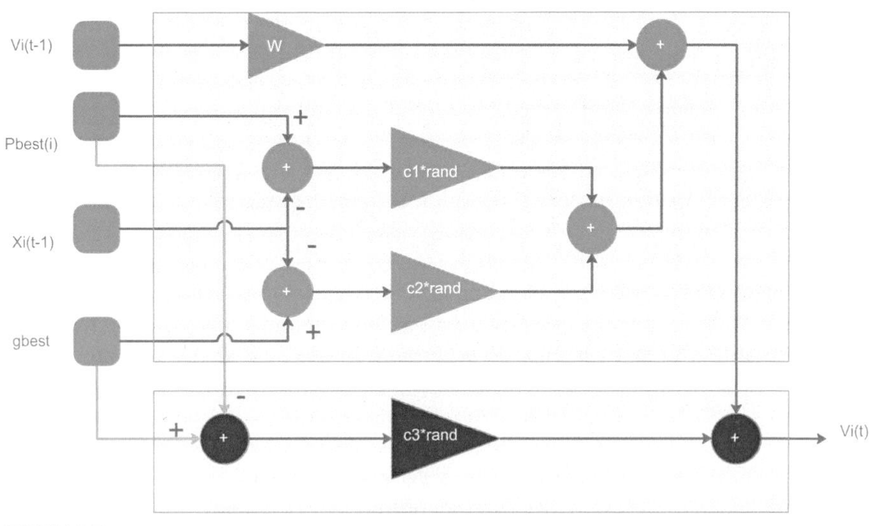

FIGURE 2.5

The architecture of the proposed method for velocity updating.

- If we are in the first iteration ($t = 1$). The current position of each particle is considered as the best location for that particle:

$$\begin{aligned} pbest(i) &= X_i(t), \quad i = 1, 2, 3, \ldots, d \\ fitness(pbest(i)) &= fitness(X_i(t)) \end{aligned} \tag{2.21}$$

- In the rest of the iterations, the value obtained for the particles in step 2 is compared with the value of the best cost obtained for each particle. If this cost is less than the best-recorded cost for this particle, then the location and cost of this particle will replace the previous one. Otherwise, there will be no change in the location and cost recorded for this particle:

$$\begin{cases} if \ \ fitness(X_i(t)) < fitness(pbest(i)) \rightarrow pbest(i) = X_i(t) \\ else \rightarrow Not \ \ change \end{cases} \tag{2.22}$$

Step 4: Updating the velocity vector of all particles.

The velocity vector is updated based on the proposed Eq. (2.18).

Step 5: Moving particles to the new positions.

By using the velocity vector, the position vector is updated and particles move toward new positions.

Step 6: Convergence test.

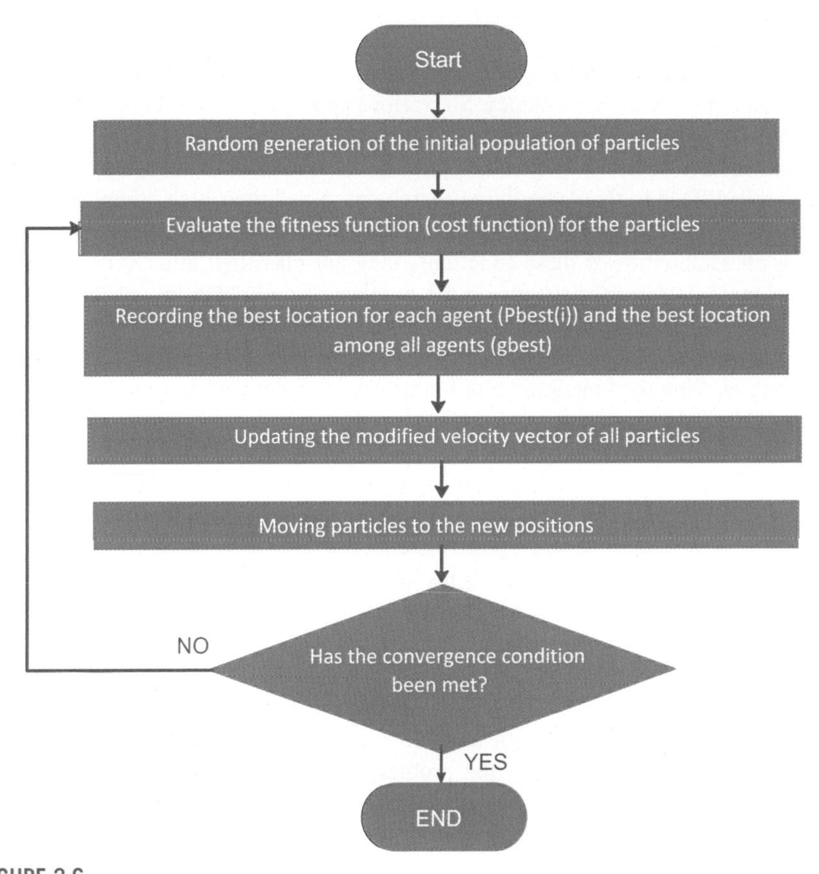

FIGURE 2.6

Block diagram of the Mo-PSO algorithm (proposed optimization algorithm).

The convergence test in this algorithm is similar to other optimization methods. There are various techniques to check the convergence of the algorithm. In this paper, a certain number of repetitions are determined. When the iteration number arrived at the determined value, the algorithm ends. Otherwise, it returns to step 2 and the rest of the steps are repeated. These steps are briefly illustrated in the flowchart in Fig. 2.6.

4. Results and discussion

This paper aimed to use the evolutionary Mo-PSO algorithm to optimize the selection of optimal features in the epileptic feature selection problem and to choose the optimal parameters of NNs in the problem of motor imagery classification.

4.1 Applying Mo-PSO algorithm on the benchmark functions

In most optimization problems, the use of analytical approaches is not efficient and does not lead to desirable accuracy. So, various approximation techniques have been suggested to solve these sets of problems with optimal approximation. Many of these techniques have obstacles such as converging in the local optimums and low convergence speed. Due to their very large search space and their high complexity, some functions are very proper for testing the capability of optimization algorithms. Based on the characteristics of these functions, they are classified into different groups. One of these categories depends on the number of optimal points of the function. The functions that have just a single optimum in the problem space are called unimodal functions and the functions that have multiple optimums in the problem search space are multimodal functions. Multimodal functions are used to evaluate the capability of the EAs to escape local optimums. If the algorithm's exploration process works poorly, it cannot fully search the entire predicament space, thus trapping in the local optimums. If the algorithm's exploration process works poorly, it cannot fully search the entire problem space, thus trapping in the local optimums. In addition, in order not to affect the reported results from the possible results, we execute the algorithm 20 courses independently on several benchmark functions and then extract the results. The list of benchmark functions used in this paper is shown in Table 2.1. In this table, the mathematical formulation, dimension, and the range of each benchmark function are reported. Among the evolutionary search algorithms, two exploration and exploitation factors are very important, so it is important to establish a balance between these factors. Exploration means that the algorithm has enough power to search in the problem space. Exploitation is the ability to get better solutions around one solution. For studying the power of the proposed algorithm in obtaining the global optimum, it is applied to the benchmark functions and is compared with other popular evolutionary techniques, such as PSO, GA, DE, and SA. The graphical results and the optimal value obtained by each algorithm are shown in Tables 2.2 and 2.3, respectively. The results provided in Table 2.2, compare the convergence of the proposed method, Mo-PSO, with others. For the Mo-PSO algorithm, the parameters are set as follows, $c_1 = c_2 = 2$, $c_3 = 1$, w changes linearly from 0.9 to 0.2, and the maximum number of repetitions is equal to 500. In addition, the stop criterion here is the maximum number of repetitions. To avoid the generalization problem, the results of 20 different independent runs have been averaged and reported. Statistically speaking, in most of these functions (Function1 to Function12), the proposed algorithm of this paper has been able to approach the global point with a very significant difference from the other evolutionarily investigated algorithms. The results reported in Tables 2.2 and 2.3 confirm this claim. In addition, in most of the convergence graphs presented in Table 2.1, it can be interpreted that the suggested method (Mo-PSO) is the best one in terms of convergences speed and reaching the optimal global point. For function 13, all of the algorithms obtained the same minimum value and in function 14, the SA algorithm is the best. In the next sections, the results of applying the proposed EA for EEG feature selection and classification are reported.

Table 2.1 Some benchmark functions to evaluate the proposed method in finding global optima.

Function number	Formula	Dimension	Range	f_{min}	Function type
FuncNo1	$f_1(x) = \sum_{i=1}^{n} x_i^2$	30	$[-100, 100]$	0	Unimodal
FuncNo2	$f_2(x) = \sum_{i=1}^{n} \lvert x_i \rvert + \prod_{i=1}^{n} \lvert x_i \rvert$	30	$[-100, 100]$	0	Unimodal
FuncNo3	$f_3(x) = \sum_{i=1}^{n} \left(\sum_{j-1}^{i} x_j \right)^2$	30	$[-100, 100]$	0	Unimodal
FuncNo4	$f_4(x) = \max\{\lvert x_j \rvert, 1 \le i \le n\}$	30	$[-100, 100]$	0	Unimodal
FuncNo5	$f_5(x) = \sum_{i=1}^{n-1} \left[100(x_{i+1} - x_i)^2 + (x_i - 1)^2 \right]$	30	$[-200, 200]$	0	Unimodal
FuncNo6	$f_6(x) = \sum_{i=1}^{n} ([x_i + 0.5])^2$	30	$[-100, 100]$	0	Unimodal
FuncNo7	$f_7(x) = \sum_{i=1}^{n} i x_i^4 + rand(0, 1)$	30	$[-1, 1]$	0	Multimodal
FuncNo8	$f_8(x) = \sum_{i=1}^{n} [x_i^2 + 10 \cos(2\pi x_i) + 10]$	30	$[-5, 5]$	0	Multimodal
FuncNo9	$f_9\left(x\right) = -20 \exp\left(-0.2 \sqrt{\frac{1}{n} \sum_{i=1}^{n} x_i^2} \right) - \exp\left(\frac{1}{n} \sum_{i=1}^{n} \cos(2\pi x_i) \right) + 20$	30	$[-20, 20]$	0	Multimodal
FuncNo10	$f_{10}(x) = \frac{1}{4000} \left(\sum_{i=1}^{n} x_i^2 \right) - \prod_{i=1}^{n} \cos\left(\frac{x_i}{\sqrt{i}} \right) + 1$	30	$[-500, 500]$	0	Unimodal

Continued

Table 2.1 Some benchmark functions to evaluate the proposed method in finding global optima.—*cont'd*

Function number	Formula	Dimension	Range	f_{min}	Function type
FuncNo11	$f_{11}(x) = \frac{\pi}{n}\left\{10\sin(\pi y_1) + \sum_{i=1}^{n-1}(y_i-1)^2\left[1+10\sin(\pi y_{i+1}+1)\right. + (y_n-1)^2\right\} + \sum_{i=1}^{n} u(x_i, 10, 100, 4)$	30	$[-10, 10]$	0	Multimodal
FuncNo12	$f_{12}(x) = 0.1\left\{\sin^2(3\pi x_1) + \sum_{i=1}^{n-1}(x_i-1)^2\left[1+\sin^2(3\pi x_i+1)\right. + (x_n-1)^2\left[1+\sin^2(2\pi x_i+1)\right]\right\} + \sum_{i=1}^{n} u(x_i, 5, 100, 4)$	30	$[-5, 5]$	0	Multimodal
FuncNo13	$f_{13}(x) = \left(\frac{1}{500} + \sum_{j=1}^{25}\frac{1}{j+\sum_{i=1}^{2}(x_i-a_i)^6}\right)^{-1}$	2	$[-100, 100]$	0	Multimodal
FuncNo14	$f_{14}(x) = \sum_{i=1}^{11}\left[a_i - \frac{x_1\left(b_i^2+b_i x_2\right)}{b_i^2+b_i x_3+x_4}\right]^2$	4	$[-5, 5]$	0	Multimodal

Table 2.2 Comparing the proposed algorithm (Mo-PSO) with other well-known EAs in finding the global minimum of the benchmark functions.

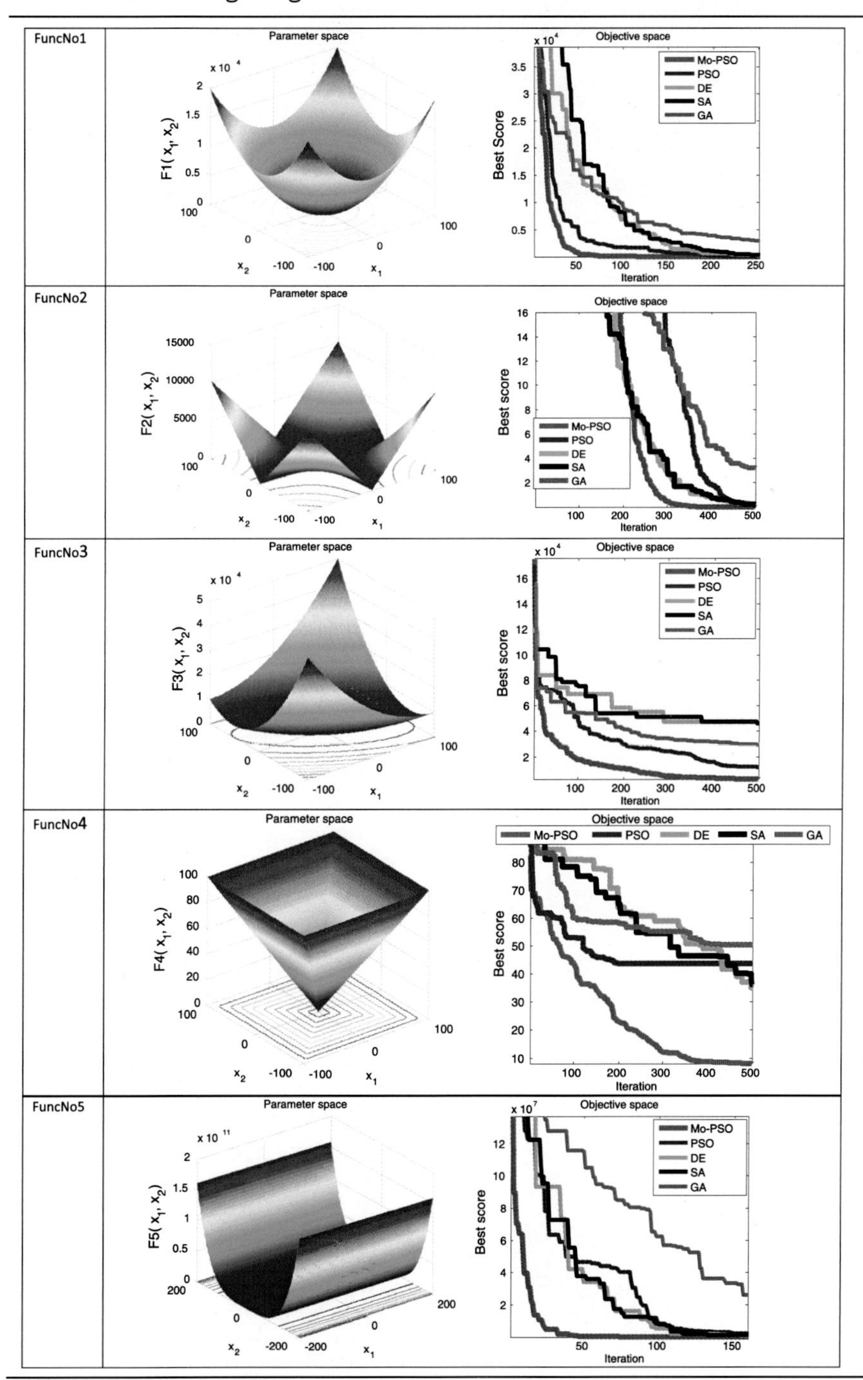

Continued

Table 2.2 Comparing the proposed algorithm (Mo-PSO) with other well-known EAs in finding the global minimum of the benchmark functions.—*cont'd*

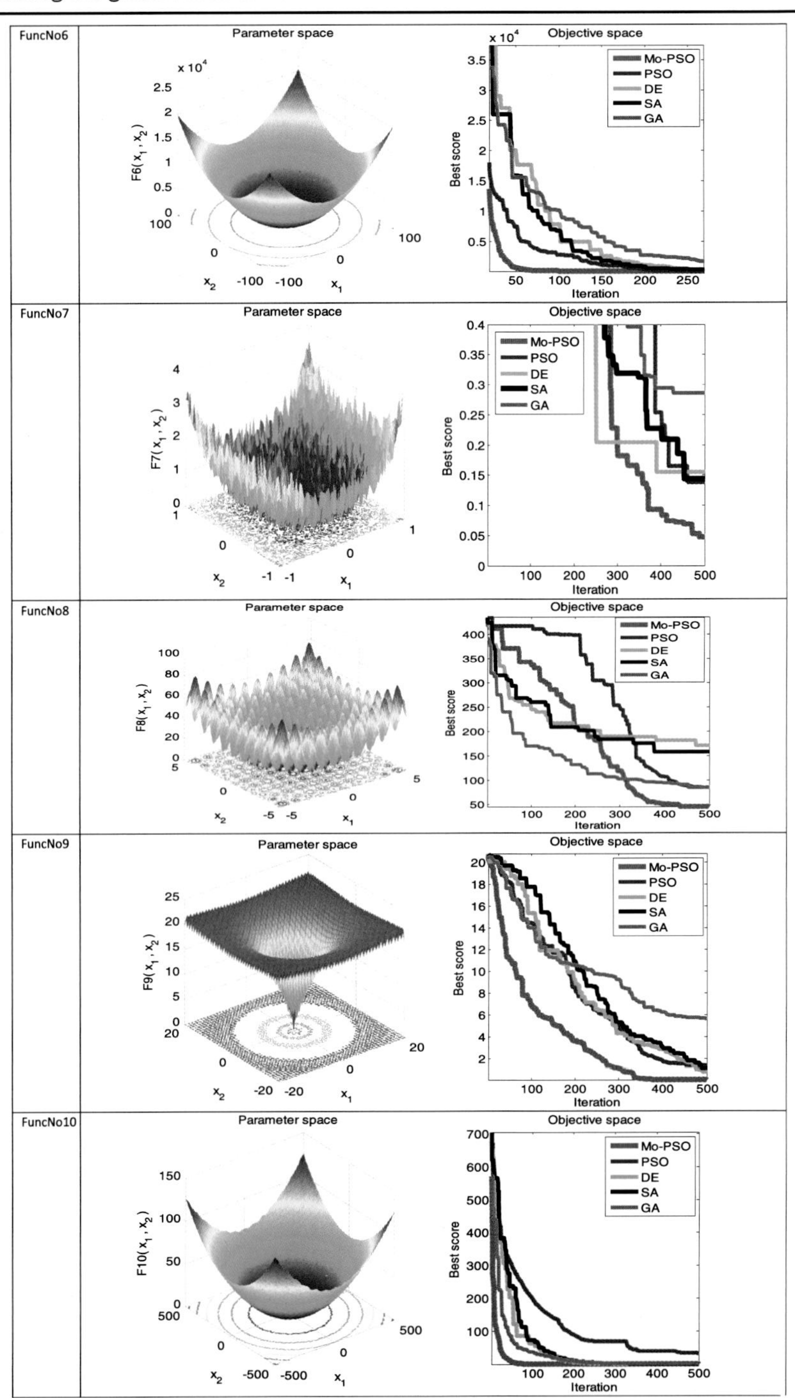

Table 2.2 Comparing the proposed algorithm (Mo-PSO) with other well-known EAs in finding the global minimum of the benchmark functions.—*cont'd*

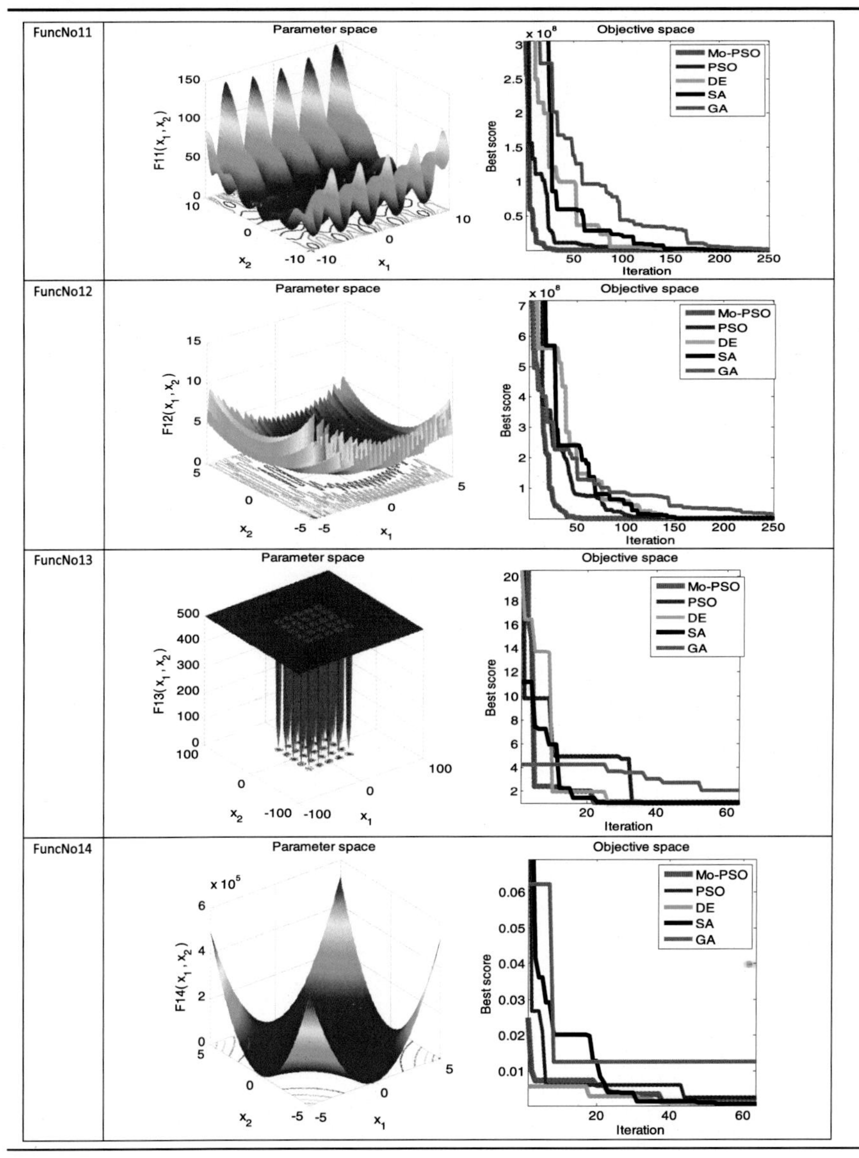

Table 2.3 Final value obtained by each algorithm.

Algorithm Function name	PSO	Mo-PSO	SA	DE	GA
FuncNo1	0.0083	**1.2763e-06**	1.9476	1.2697	557.6018
FuncNo2	0.2197	**8.7466e-04**	0.2031	0.2432	3.2043
FuncNo3	1.2078e+04	**2.5014e+03**	4.6254e+04	4.7797e+04	2.9497e+04
FuncNo4	43.7355	**8.0379**	36.3461	35.1277	50.3761
FuncNo5	410.6539	**31.1475**	986.3280	1.4344e+03	5.4037e+05
FuncNo6	0.9645	**3.7081e-07**	1.6825	1.4859	581.3712
FuncNo7	0.1387	**0.0477**	0.1442	0.1552	0.2857
FuncNo8	84.8873	**44.8005**	158.5592	171.6066	84.7915
FuncNo9	1.3522	**0.0038**	1.0976	0.7531	5.6436
FuncNo10	33.8715	**0.0345**	1.0304	0.9680	2.0218
FuncNo11	4.6821	**8.9153e-08**	5.4089	3.6954	343.2275
FuncNo12	2.0013e+06	**6.3925e-07**	6.3041	19.9038	1.1622e+04
FuncNo13	0.9980	0.9980	0.9980	0.9980	0.9980
FuncNo14	9.8098e-04	8.4222e-04	**3.0846e-04**	5.1264e-04	0.0046

4.2 The results of applying Mo-PSO for epilepsy feature selection

One of the applications examined in this article was the selection of optimal features for an epilepsy diagnosis. In fact, we claimed that the fewer selected features, the less complexity and power consumption of the diagnostic system. In this paper, we suggest the Mo-PSO technique to select the optimal features that bring together the data from these optimal features and provide the optimal dataset for an epilepsy diagnosis. The art of the EA is to combine the data of one set of electrodes in each test and, by minimizing the fitness function, obtain the combination of these electrodes that leads to the best fitness value.

For a comprehensive comparison, the suggested method is compared with other well-known methods in the field of EAs, and the results are shown in Table 2.4 and Fig. 2.7. As can be observed, the number of selected electrodes is specified as the input parameter of these algorithms, and the user determines how many of these electrodes are selected and activated. For this purpose, the number of selected electrodes has been changed from 70 to 20 and for each of this number of selected electrodes, the corresponding accuracy has been obtained. Here, to avoid generalization problems, each test was run 20 times and their averages were reported. The lower the number of features, the lower the accuracy. However, one algorithm can select some electrodes in terms of fitness value evaluation and consider that set as the feature set and exclude the other electrodes that contain redundant information. An algorithm works well to obtain the desired accuracy by selecting a few electrodes, which decrease the cost of the detection system and reduce the complexity of the system significantly. The results in Table 2.4 show how much the reduction in the accuracy was due to the decrease in the number of features. In addition, in Fig. 2.7, the slope of the diagrams of each algorithm can be a criterion for evaluating the effect of electrode reduction on detection accuracy.

The lower slope reflects the fact that the algorithm is well able to select the electrodes that contain the optimal data and ignore the rest of the electrodes. With these interpretations, the results confirm that the suggested algorithm can efficiently select the optimal electrode. As shown in Table 2.4, in test 10, where the lowest number of electrodes (20 electrodes) was investigated, the accuracies for SA, DE, and PSO algorithms were 60.2019%, 52.5991%, and 57.2159%, respectively, which are not desirable values for a diagnostic system. However, under these conditions, the accuracy of GA, with 20 electrodes, reached 82.1361%. Among all, the proposed algorithm outperforms the rest of the algorithms, with only 20 electrodes obtained an accuracy of 89.2597%. Therefore, the use of low electrodes and acceptable accuracy have been achieved using the Mo-PSO algorithm and NNs. In addition, to demonstrate the efficiency of the method and compare better, comparisons with other algorithms such as sequential feature selection [30], conditional mutual information [31], Plus-L Minus-R selection [32], and sequential backward elimination [33] have been performed. The results of this comparison are provided in Fig. 2.8. As can be seen, the proposed algorithm outperforms the other algorithms in terms of classification rate and a number of selected electrodes.

Table 2.4 The results of the epilepsy feature selection test using the proposed method (Mo-PSO) and other EAs.

Algorithm: PSO+NNs

Test Number	Accuracy (%)	NSF
TestNo1	86.1205	70
TestNo2	81.7560	65
TestNo3	73.5644	60
TestNo4	69.9456	50
TestNo5	65.6875	45
TestNo6	64.7641	40
TestNo7	62.3058	35
TestNo8	59.2598	30
TestNo9	58.6581	25
TestNo10	57.2159	20

Algorithm: GA+NNs

Test Number	Accuracy (%)	NSF
TestNo1	94.0025	70
TestNo2	92.1258	65
TestNo3	91.3951	60
TestNo4	89.7415	50
TestNo5	88.0178	45
TestNo6	85.1219	40
TestNo7	85.0289	35
TestNo8	84.2569	30
TestNo9	83.8790	25
TestNo10	82.1361	20

Algorithm: DE+NNs

Test Number	Accuracy (%)	NSF
TestNo1	85.1059	70
TestNo2	80.1256	65
TestNo3	78.8569	60
TestNo4	75.1599	50
TestNo5	74.3692	45
TestNo6	65.1258	40
TestNo7	63.7895	35
TestNo8	58.3682	30
TestNo9	55.8036	25
TestNo10	52.5991	20

Algorithm: SA+NNs

Test Number	Accuracy (%)	NSF
TestNo1	84.5581	70
TestNo2	83.0258	65
TestNo3	81.2598	60
TestNo4	79.2568	50
TestNo5	78.2541	45
TestNo6	75.6958	40
TestNo7	70.8103	35
TestNo8	65.5991	30
TestNo9	63.2897	25
TestNo10	60.2019	20

Algorithm: **Mo-PSO+NNs** (proposed)

Test Number	Accuracy (%)	NSF
TestNo1	97.7995	70
TestNo2	96.1289	65
TestNo3	96.0259	60
TestNo4	95.3356	50
TestNo5	94.5964	45
TestNo6	94.1789	40
TestNo7	93.3587	35
TestNo8	92.0459	30
TestNo9	91.9931	25
TestNo10	89.2597	20

4.3 The results of using Mo-PSO for motor imagery classification

Another application discussed in this paper is the motor imagery classification. In this paper, as it was presented in the feature extraction phase, the CSP method was used for this application. The CSP method is very useful in two-class diagnostic systems. In other words, using this method, the data are mapped to the new space, which will increase the distance between the two classes and is suitable for

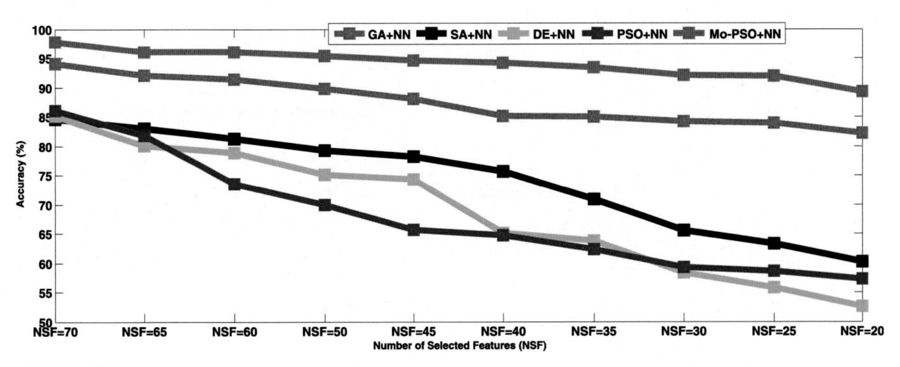

FIGURE 2.7

Comparison of accuracies related to each feature set.

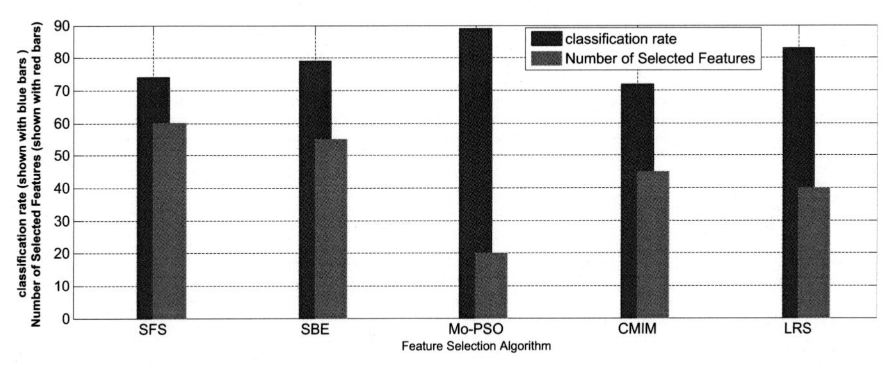

FIGURE 2.8

Comparing proposed algorithms with other feature selection algorithms.

classification. Another important point about this algorithm is that in addition to changing the data space, it also performs the feature reduction. Therefore, by selecting the appropriate value of the parameter m, the data in the new space have 2*m features. In our simulations, the value of this parameter is $m = 5$, so we finally have 10 features. If we can design a diagnostic system with such a small number of features, the system will be greatly optimized in terms of efficiency and complexity. We first applied these features to well-known classifiers such as NNs [34], SVM [35], KNN [36], and LDA [37], which yielded the accuracies of equal to 77.3562%, 74.1089%, 70.5796, and 80.5693, respectively. However, these are not impressive. In this paper, we use NNs whose parameters were optimized using the proposed EA. As described in the methods section, we considered the optimization parameters as connection weights. The art of using the proposed method in the NN training is to gain weights that minimize the fitness function. With this technique and after obtaining an optimal structure for the NN, the extracted features were applied to it. The convergence curves for 500 iterations are shown in Fig. 2.9. The simulations for seven subjects are shown in this figure. As shown, the suggested

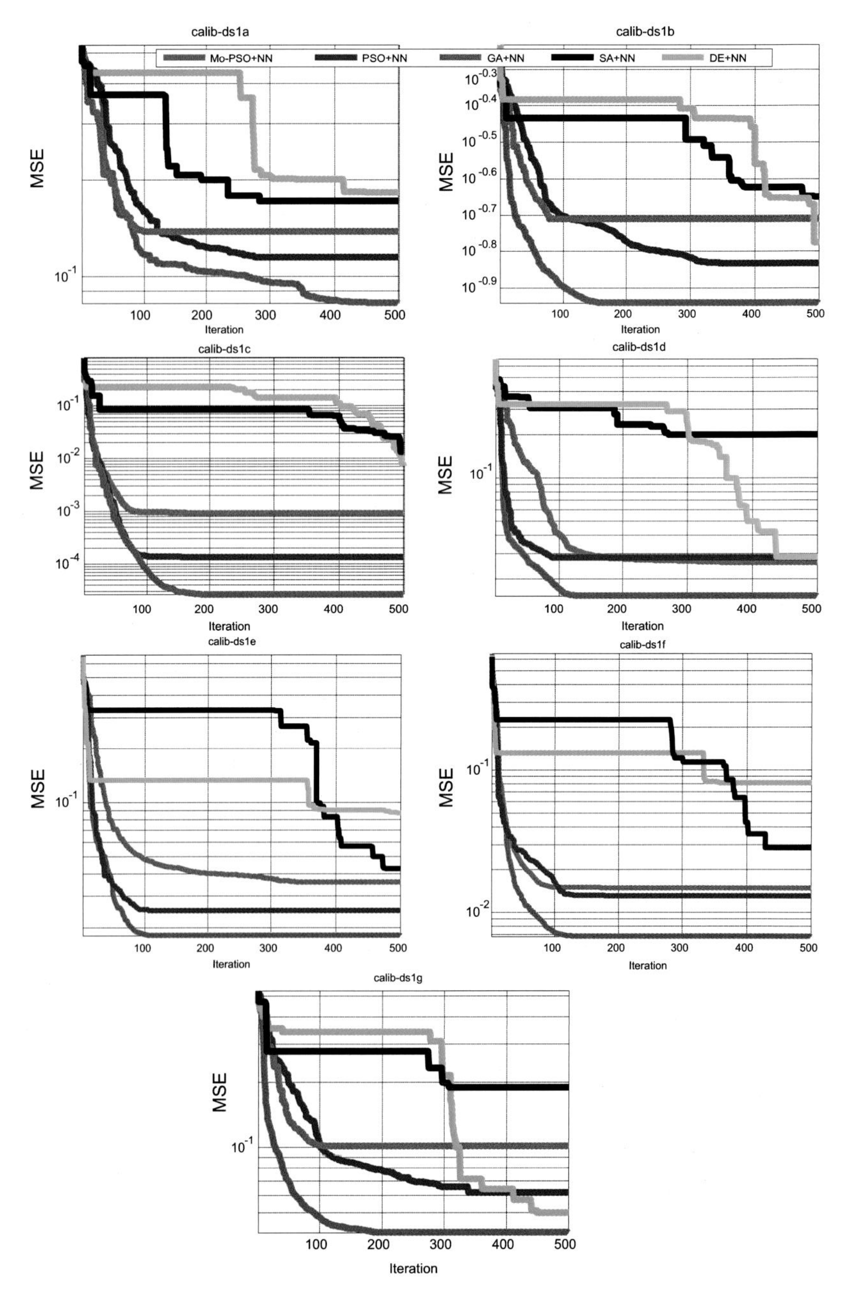

FIGURE 2.9

Convergence curves of EAs on motor imagery EEG dataset.

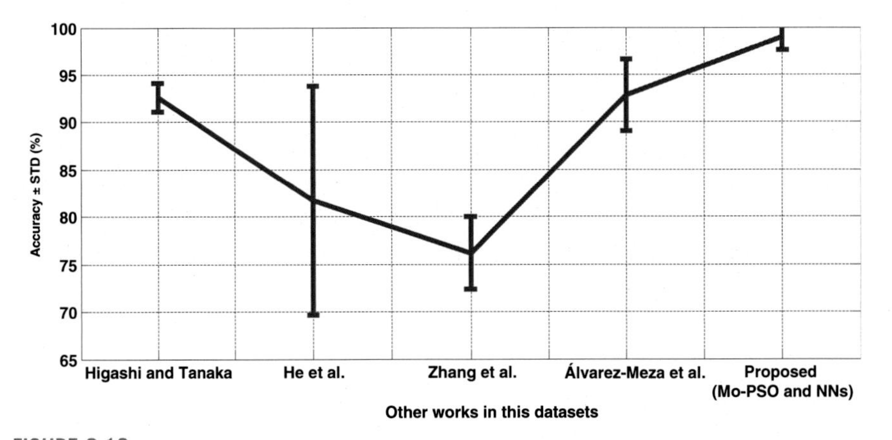

FIGURE 2.10

Comparison with other works on motor imagery classification dataset.

technique has been capable to perform the classification procedure with high accuracy and is better than the rest of the evolutionary methods. Another point to understand from these convergence curves is the convergence speed. With a little scrutiny of the results presented, it can be found that the suggested method has been able to converge at an acceptable speed and has converged accurately in relatively low iterations. In addition, a comparison has been made with other works in Refs. [38–41] on this dataset. The results of this comparison are provided in Fig. 2.10. It can be found that the suggested technique is better than the other methods in terms of classification accuracy.

5. Conclusion

This paper provides a new version of the PSO algorithm called Mo-PSO. The idea behind this algorithm is to compare the local and global experiences for velocity updating. With this change, the exploration ability of the PSO algorithm has been enhanced. To demonstrate the efficiency of this algorithm in achieving global optimum and convergence speed, some simulations were performed on 14 benchmark functions. The simulation results revealed that the suggested approach performs more beneficial than the rest of the popular algorithms in terms of approaching the global optimum and convergence speed. After demonstrating the success of the suggested algorithm when applied to benchmark functions, it was used to select epileptic features and NN training to classify motor imagery. The feature selection is an important step so that the smaller the number of electrodes, the lower the complexity of the system. As can be seen from the results, the proposed method was able to achieve acceptable accuracy by selecting about 20% of the total electrodes. Therefore, a reasonable compromise can be made between the accuracy and the number of selected electrodes. Another application discussed in this paper was the motor imagery classification. First, 10 features were extracted using the CSP method. In this work, the proposed algorithm was used in the classification phase. The proposed algorithm has been used to train NNs and gain optimal weights. After the parameter setting, the extracted features were applied to this classifier. The results revealed that the suggested approach has better or comparable performance than other methods. For future work, it is suggested that by inspiring nature, a new algorithm will be designed and used to optimize the problems and make them appear in real applications.

References

[1] Goldenberg MM. Overview of drugs used for epilepsy and seizures: etiology, diagnosis, and treatment. Pharmacy and Therapeutics 2010;35(7):392. https://www.ncbi.nlm.nih.gov/pubmed/20689626.

[2] Srinivasan V, Eswaran C, Sriraam N. Approximate entropy-based epileptic EEG detection using artificial neural networks. IEEE Transactions on Information Technology in Biomedicine 2007;11(3):288–95.

[3] Ocak H. Automatic detection of epileptic seizures in EEG using discrete wavelet transform and approximate entropy. Expert Systems with Applications 2009;36(2): 2027–36.

[4] Tzallas AT, Tsipouras MG, Fotiadis DI. Epileptic seizure detection in EEGs using time–frequency analysis. IEEE Transactions on Information Technology in Biomedicine 2009;13(5):703–10.

[5] Acharya UR, Molinari F, Sree SV, Chattopadhyay S, Ng K-H, Suri JS. Automated diagnosis of epileptic EEG using entropies. Biomedical Signal Processing and Control 2012;7(4):401–8.

[6] Subasi A. EEG signal classification using wavelet feature extraction and a mixture of expert model. Expert Systems with Applications 2007;32(4):1084–93.

[7] Subasi A, Gursoy MI. EEG signal classification using PCA, ICA, LDA and support vector machines. Expert Systems with Applications 2010;37(12):8659–66.

[8] Martis RJ, et al. Application of empirical mode decomposition (EMD) for automated detection of epilepsy using EEG signals. International Journal of Neural Systems 2012;22(6):1250027.

[9] Orhan U, Hekim M, Ozer M. EEG signals classification using the K-means clustering and a multilayer perceptron neural network model. Expert Systems with Applications 2011;38(10):13475–81.

[10] Sharma R, Pachori RB. Classification of epileptic seizures in EEG signals based on phase space representation of intrinsic mode functions. Expert Systems with Applications 2015;42(3):1106–17.

[11] Patidar S, Panigrahi T. Detection of epileptic seizure using Kraskov entropy applied on tunable-Q wavelet transform of EEG signals. Biomedical Signal Processing and Control 2017;34:74–80.

[12] Fu K, Qu J, Chai Y, Zou T. Hilbert marginal spectrum analysis for automatic seizure detection in EEG signals. Biomedical Signal Processing and Control 2015;18: 179–85.

[13] Arunkumar N, et al. Classification of focal and non focal EEG using entropies. Pattern Recognition Letters 2017;94:112–7.
[14] Schalk G, McFarland DJ, Hinterberger T, Birbaumer N, Wolpaw JR. BCI2000: a general-purpose brain-computer interface (BCI) system. IEEE Transactions on Biomedical Engineering 2004;51(6):1034–43.
[15] Qin L, Ding L, He B. Motor imagery classification by means of source analysis for brain–computer interface applications. Journal of Neural Engineering 2004;1(3):135.
[16] Afrakhteh S, Mosavi MR, Khishe M, Ayatollahi A. Accurate classification of EEG signals using neural networks trained by hybrid population-physic-based algorithm. International Journal of Automation and Computing 2018. https://link.springer.com/article/10.1007/s11633-018-1158-3.
[17] Mosavi MR, Ayatollahi A, Afrakhteh S. An efficient method for classifying motor imagery using CPSO-trained ANFIS prediction. Evolving Systems 2019:1–18. https://link.springer.com/article/10.1007/s12530-019-09280-x.
[18] Eberhart R, Kennedy J. A new optimizer using particle swarm theory. In: Proceedings of the Sixth International Symposium on Micro machine and Human science; 1995. p. 39–43. https://doi.org/10.1109/MHS.1995.494215. 1995.
[19] Van Laarhoven PJM, Aarts EHL. Simulated annealing. In: Simulated annealing: theory and applications. Springer; 1987. p. 7–15. https://link.springer.com/chapter/10.1007/978-94-015-7744-1_2.
[20] Storn R, Price K. Differential evolution–a simple and efficient heuristic for global optimization over continuous spaces. Journal of Global Optimization 1997;11(4):341–59. https://link.springer.com/article/10.1023/A:1008202821328.
[21] Whitley D. A genetic algorithm tutorial. Statistics and Computing 1994;4(2):65–85. https://link.springer.com/article/10.1007/BF00175354.
[22] http://epileptologie-bonn.de/cms/front_content.php?idcat=193&lang=3.
[23] Selesnick IW, Burrus CS. Generalized digital Butterworth filter design. IEEE Transactions on Signal Processing 1998;46(6):1688–94.
[24] Blankertz B, Dornhege G, Krauledat M, Müller KR, Curio G. The non-invasive Berlin brain–computer interface: fast acquisition of effective performance in untrained subjects. NeuroImage 2007;37(2):539–50.
[25] Ramoser H, Muller-Gerking J, Pfurtscheller G. Optimal spatial filtering of single trial EEG during imagined hand movement. IEEE Transactions on Rehabilitation Engineering 2000;8(4):441–6.
[26] McCulloch WS, Pitts W. A logical calculus of the ideas immanent in nervous activity. Bulletin of Mathematical Biophysics 1943;5(4):115–33. https://link.springer.com/article/10.1007/BF02478259.
[27] Rosenblatt F. Principles of neurodynamics. Perceptrons and the theory of brain mechanisms. Buffalo NY: Cornell Aeronautical Lab Inc; 1961. https://apps.dtic.mil/docs/citations/AD0256582.
[28] Lagaris IE, Likas AC, Papageorgiou DG. Neural-network methods for boundary value problems with irregular boundaries. IEEE Transactions on Neural Networks 2000;11(5):1041–9.
[29] Rumelhart DE, Hinton GE, Williams RJ. Learning representations by back-propagating errors. Cognitive Modeling 1988;5(3):1.
[30] Rückstieß T, Osendorfer C, van der Smagt P. Sequential feature selection for classification. In: Australasian joint conference on artificial intelligence; 2011. p. 132–41. http://www.doi.org/10.1007/978-3-642-25832-9_14.

[31] Fleuret F. Fast binary feature selection with conditional mutual information. Journal of Machine Learning Research 2004;5:1531–55. http://www.jmlr.org/papers/v5/fleuret04a.html.

[32] Somol P, Pudil P, Novovičová J, Paclık P. Adaptive floating search methods in feature selection. Pattern Recognition Letters 1999;20(11–13):1157–63.

[33] Mao KZ. Orthogonal forward selection and backward elimination algorithms for feature subset selection. IEEE Transactions on Systems, Man, and Cybernetics, Part B (Cybernetics) 2004;34(1):629–34.

[34] Hansen LK, Salamon P. Neural network ensembles. IEEE Transactions on Pattern Analysis and Machine Intelligence 1990;12(No. 10):993–1001.

[35] Cortes C, Vapnik V. Support-vector networks. Machine Learning 1995;20(No.3): 273–97. https://rd.springer.com/article/10.1007%2FBF00994018.

[36] Altman NS. An introduction to the Kernel and nearest neighbor nonparametric regression. The American Statistician 1992;46(No.3):175–85.

[37] Pang S, Ozawa S, Kasabov N. Incremental linear discriminant analysis for classification of data streams. IEEE transactions on Systems, Man, and Cybernetics, part B (Cybernetics) 2005;35(5):905–14.

[38] Higashi H, Tanaka T. Common spatio-time-frequency patterns for motor imagery-based brain-machine interfaces. Computational Intelligence and Neuroscience 2013;2013: 1–13.

[39] He W, Wei P, Wang L, Zou Y. A novel EMD-based common spatial pattern for motor imagery brain-computer interface. IEEE-EMBS International Conference on Biomedical and Health Informatics 2012:216–9.

[40] Zhang H, Guan C, Ang KK, Wang C, Chin ZY. BCI competition IV – data set I: learning discriminative patterns for self-paced EEG-based motor imagery detection. Frontiers in Neuroscience 2012;6(No.FEB):1–7.

[41] Álvarez-Meza AM, Velásquez-Martínez LF, Castellanos-Dominguez G. Time-series discrimination using feature relevance analysis in motor imagery classification. Neurocomputing 2015;151:122–9.

CHAPTER

3 Edge computing for energy-efficient smart health systems: data and application-specific approaches

Alaa Awad Abdellatif[1,2], Amr Mohamed, PhD[3], Carla Fabiana Chiasserini[2], Aiman Erbad, PhD[1], Mohsen Guizani[3]

[1]*Department of Computer Science and Engineering, Qatar University, Doha, Qatar;* [2]*Department of Electronics and Telecommunications, Politecnico di Torino, Torino, Italy;* [3]*Professor, Department of Computer Science and Engineering, Qatar University, Doha, Qatar*

1. Introduction

Healthcare has gained a significant interest all over the world because of its importance in promoting human development, and the well-being of countries' citizens. The growing number of patients with chronic disease, disaster management, and emerging epidemiological threats pose great challenges for governments and public sectors. They motivate such entities to welcome, adopt, and support the development of increasingly innovative healthcare approaches and initiatives [1]. However, traditional healthcare systems cannot support the scalability required to meet the rising number of patients as they require one-to-one relationships between the caregiver and the patient. One of the key concepts for mitigating healthcare scalability is to have patients participate in their own treatment by providing them with intuitive, nonintrusive tools that allow them to efficiently communicate with their caregivers.

The rapid development of intelligent systems and Wearable Internet of Things (WIoT) devices, in addition to the advances in mobile communication technologies, have fostered the evolution of traditional healthcare systems into smart health systems. At the beginning, the concept of remote health, also called as tele-health, has been appeared as a new concept where patients and/or caregivers would be able to utilize mobile technologies to remotely deliver health information. This could potentially help reduce hospitalization and deliver timely healthcare to remote societies at low cost [2]. Then, mobile-Health (m-Health) systems have manifested to provide new ways of acquiring, processing, and transferring processed data to deliver meaningful results.

Smart-health (s-Health) represents the context-aware development of m-Health, exploiting communication technology to equip healthcare stakeholders with

Energy Efficiency of Medical Devices and Healthcare Applications. https://doi.org/10.1016/B978-0-12-819045-6.00003-0

innovative solutions and tools that can revolutionize healthcare industry. s-Health systems comprise various wireless medical devices, sensors, cameras, and WIoT devices that play a significant role in real-time biosignal monitoring, enabling automatic tracking of the patients, and controlling patients' drugs usage. Hence, they allow for early detection of clinical deterioration, such as seizure detection and heart failure. However, all these devices generate an enormous amount of information that require processing, readily transferring, and storing, while maintaining security and privacy protection. Such requirements turn the classic cloud computing framework inadequate for s-Health, because the centralized management of such amount of data cannot provide the required level of scalability and high responsiveness needed for s-Health applications.

Accordingly, mobile or multiaccess edge computing (MEC) has recently emerged to provide the capabilities needed for processing and managing the acquired data at the proximity of the data sources (i.e., at the network edge) [3,4]. Thus, given the aforementioned characteristics and requirements of s-Health, we envision that edge computing can significantly benefit the healthcare evolution to smart healthcare through enabling better insight of heterogeneous healthcare media content to provide affordable and high-quality patient care. Edge computing along with the next-generation networking technologies can be the technical-driven factors for realizing the vision of smart healthcare services as they will accelerate data generation and processing, while allowing the resource constrained devices to communicate efficiently with the healthcare stakeholders. In particular, the main benefits of MEC in a smart-health environment can be highlighted as follows:

1. Enabling short response time and fast emergency prediction and detection response.
2. Decreasing power consumption for battery-operated IoT devices.
3. Optimizing network bandwidth utilization.
4. Providing secure medical data transmission and privacy protection.

This chapter presents an edge-based s-Health system architecture for reliable, scalable, and effective patient monitoring. The proposed architecture leverages sensors and wireless networking technologies for connecting patients with medical healthcare providers to enable early diagnosis, remote monitoring, and fast emergency response for the elderly and chronic disease patients. In contrast to the previous work in this domain, the adopted framework considers context-aware approaches by focusing on applications' requirements and patients' data characteristics, leveraging heterogeneous wireless networks for optimizing medical data delivery. Accordingly, we focus in this chapter on answering the following questions:

1. How to decrease transmitted data size, while maintaining reliable real-time healthcare services?
2. How to incorporate wireless network components with application's characteristics to develop energy-efficient s-Health system?
3. How to utilize the spectrum across multiple radio access technologies to fulfill applications' QoS?

In this chapter, Section 2 presents the proposed MEC-based system architecture that satisfies the s-Health requirements, highlighting the advantages of implementing intelligent data processing techniques at the network edge. Section 3 introduces some of these edge computing techniques including adaptive in-network compression, event detection, and network-aware optimization, which enable MEC-based system architecture to fulfill all s-Health requirements. Section 4 then discusses the challenges and open issues for utilizing MEC paradigm in s-Health, including the use of cooperative edges for improved energy and spectrum efficiency, as well as the need and benefit of combining heterogeneous data sources at the edge. Finally, Section 5 concludes the chapter.

2. Smart health system

This section introduces a brief description of the proposed s-Health architecture and investigates the benefits of incorporating the MEC within s-Health system.

2.1 s-Health system architecture

The proposed architecture in Fig. 3.1 considering the end-to-end healthcare system starting from the data sources attached or near to patients till ending with the healthcare providers. It includes the following main components:

Hybrid monitoring devices: It represents the set of data sources located on or around the patients for continuous monitoring of the patient's state. These sensing sources may include medical/nonmedical devices, such as implantable or wearable sensors, smartphones, and digital cameras. These hybrid sources of information are utilized within the automated smart environment for enabling continuous-remote monitoring and fast prediction/detection of emergency circumstances. Such IoT devices can be connected either with a mobile/infrastructure edge node, to process the acquired data locally, or directly with the network infrastructure (see Fig. 3.1).

Mobile or infrastructure edge: Here, we refer to the mobile edge node as a patient data aggregator (PDA) that implements the in-network processing mechanisms before forwarding the data to the cloud.

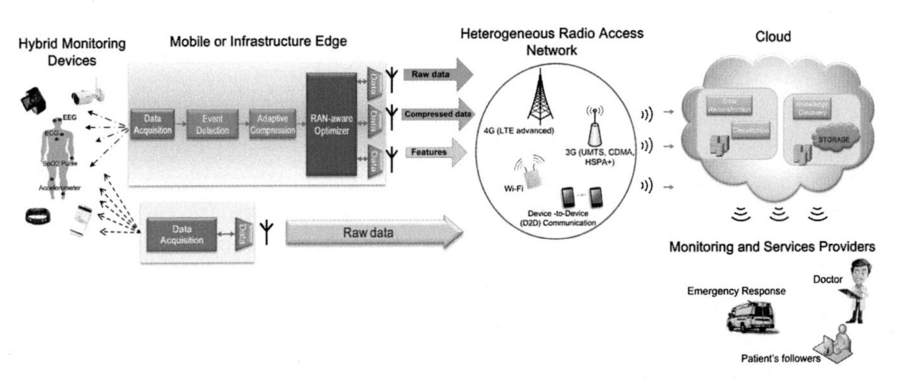

FIGURE 3.1

Proposed s-Health system architecture.

The PDA can be a smartphone that fuses the medical and nonmedical data from various monitoring devices, executes in-network processing on the acquired data, event detection, and emergency notification, and transfers the important data or extracted features of interest to the cloud. Furthermore, the PDA can be a data source itself, which generates information related to the patient's conditions. Interestingly, different health-related applications (apps) can be developed at the PDA level for enabling patient—doctors' interactions or facilitating chronic disease management. Moreover, these apps allow the patients to get involved in their treatment while interacting with their doctors anywhere and anytime. In addition to that, with a PDA running optimized context-aware processing, different monitoring devices can be managed easily at the proximity of the patient, while optimizing medical data delivery considering the environment context, that is, data characteristics, applications' requirements, and wireless network conditions.

Heterogeneous radio access network: As mentioned before, providing high-quality s-Health services results in generating enormous amount of data, which demands for high data rates. To maintain this while providing high quality of service (QoS) for s-Health, we opt to exploiting the heterogeneity of wireless network. Heterogeneous networks (HetNets) can satisfy the rising traffic demand and successfully maintain the application's QoS requirements through leveraging the availability of several technologies, such as Wi-Fi, UMTS, LTE, and Bluetooth. Hence, it enables the association with the most appropriate radio technology with the best energy consumption and data rate.

Cloud: It represents the central storage and control unit, where data storage, epidemiological threats detection, population health management, and sophisticated data analysis techniques can be implemented. Central hospital can play the role of the cloud, where data collection and patients' records analysis can be implemented to provide the needed assistance.

Monitoring and healthcare service providers: Healthcare service providers can be doctors, ambulances, or even a patients' relatives, who provide curative, rehabilitative, or emergency services to the patients.

2.2 Advantages of s-Health

In the light of the aforementioned characteristics and requirements of s-Health system, the advantages of the proposed system architecture mentioned earlier can be summarized as follows:

Data reduction: Various sensors, cameras, and medical devices utilized in s-Health systems continuously generate a massive amount of data every few seconds [2]. For instance, electroencephalogram (EEG) monitoring applications typically use high-resolution headset devices containing up to 100 electrodes, each generating data with sampling rate around 1000 samples/second, which leads to a data rate of 1.6 Mbps per single device per single patient. Hence, using centralized cloud paradigm to support such traffic demand is not advisable and may turn some of the s-Health services to be impractical, given the limited radio resources. Accordingly, applying advanced edge-based processing techniques at the collected data can significantly decrease the amount of transmitted data toward the cloud, hence enhancing energy efficiency and bandwidth consumption.

Energy efficiency: s-Health systems are usually composed of diverse IoT devices that require to be used for a long time before replacement. Thus, continuous data transmission is not possible because of the high energy consumption it causes. Optimizing the devices' operational states and their data transmission at the edge facilitates a better usage of devices' batteries, in addition to the proximity between these devices and the edge, which further decreases the energy consumption resulting from data transmission (a component that is estimated for example for a wireless EEG monitoring system by 70% of the total energy consumption [5]). Accordingly, leveraging adaptive data compression and selection of the most convenient radio interface at the edge for data transmission toward the cloud can significantly reduce the energy consumption.

Swift response: For real-time monitoring applications, only main information about patients' states can be reported to the cloud, in normal health conditions, with loose delay constraint; whereas, in the case of emergency, the swift delivery of intensive amount of data to the cloud is a necessity. To achieve that, data are required to be analyzed and even a diagnosis is made as close as possible to the patient. The proposed s-Health system can address this issue using the ability of the edge node (PDA) to execute event-detection techniques to detect the emergency conditions.

Location awareness: The edge node can be fruitfully leveraged to infer important context information that is used for localization methods. This brings two main benefits to s-Health system. First, localizing a patient facilitates matching his/her geographical location with the nearest caregiver, for example, hospital or ambulance. Second, data transfer can be optimized taking into consideration the nearest mobile edge node, or the most convenient device that can forward the data to the cloud, which ultimately improves energy efficiency and reliability.

3. Possible approaches to be implemented at the edge

This section demonstrates the main functionality that can be implemented at the edge. Specifically, different context-aware approaches are presented to optimize medical data delivery and QoS for s-Health by moving computational intelligence to the network edge. These approaches include data-specific technique, which considers data characteristics such as sparsity to adaptively adjust transmitted data size based on application's requirements, and state of the wireless network; application-specific technique, which uses characteristics related to the application such as class of the data to obtain transmitted data type; and network-aware technique that allows the PDA to be connected anywhere and anytime, while optimizing network association.

3.1 Adaptive in-network compression

The classic approach of transferring the entire raw data wirelessly to the cloud requires the transmission of an enormous amount of data, which is challenging. A promising methodology to address this challenge in s-Health system is to perform local in-network processing and data-specific compression on the collected data considering the network state before the transmission. This facilitates, on one hand, implementing efficient compression techniques with high compression ratio and low signal distortion, on the other hand, decreasing transmitted data size, hence decreasing transmission energy.

A possible approach to tackle this problem is designing a holistic energy—cost—distortion framework. This framework leverages the benefits of adaptive compression to optimize not only transmission energy consumption, but also to account for monetary cost of using network services as well as the requirements on signal distortion for medical data. In particular, this approach formulates a multiobjective optimization framework that accounts for minimizing the transmission energy consumption (at the physical layer), as well as the signal distortion and network utilization cost (at the application layer) through obtaining the optimal transmission rate and compression ratio, while maintaining latency and bit error rate (BER) constraints [6]. Thus, a PDA can adapt its transmission parameters according to wireless channel conditions and application's characteristics. Thus, the following tasks are implemented at the PDA:

- Receiving from sensor nodes the acquired data, and application layer constraints, for example, maximum BER and latency.
- Given the wireless network conditions, finding optimal transmission rate and compression ratio that provides the optimal trade-off among its objectives (i.e., energy consumption, monetary cost, and signal distortion).
- Compressing collected data.
- Forwarding compressed data to the cloud.

Implementing such adaptive compression schemes often leads to a trade-off between energy and distortion: the higher the compression ratio, the lower the energy consumption and the higher the distortion. This trade-off is demonstrated in Fig. 3.2.

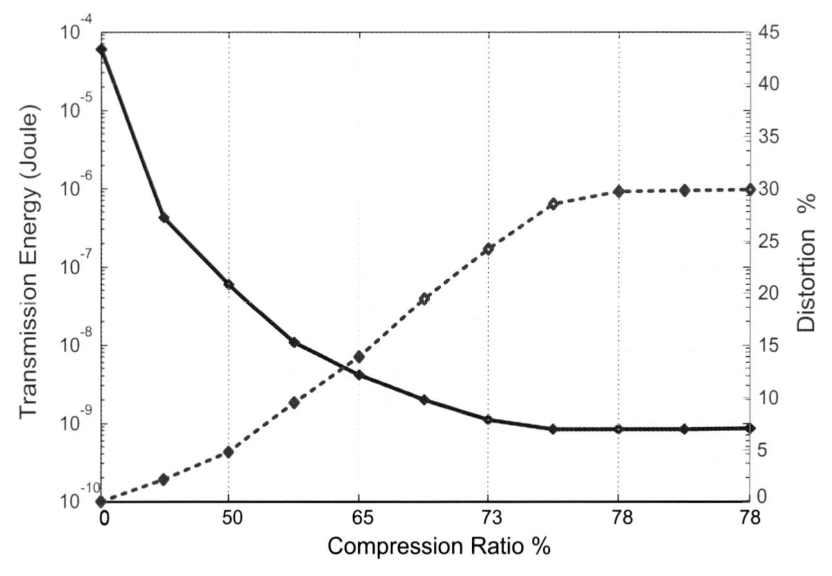

FIGURE 3.2

Trade-off between transmission energy and distortion using adaptive compression.

As shown, at low compression ratio, the obtained distortion is low, while the transmission energy is high. As reducing transmission energy becomes more important, that is, compression ratio is increased, the obtained distortion increases until it reaches a maximum target value (i.e., set at 30%), at the expense of reducing transmission energy. This result proposes that it is important to develop an algorithm that maintains the optimal trade-off among transmission energy and distortion, such that the obtained minimum value of transmission energy allows the system to satisfy the required maximum level of distortion accepted by the application.

A number of solutions have been also proposed in the literature targeting the reduction of energy consumption in body sensor networks (BSN). The main aim of these solutions varies, ranging from lossiness and computational complexity reduction to the exploitation of spatial or temporal redundancy and of waveform transformations (e.g., vector quantization and discrete wavelet transform) [7]. Specifically, two main data reduction approaches have been investigated: compressive sensing (CS) and feature extraction. The application of CS in BSN has exhibited great promise. The idea of CS is to utilize the sparsity of the input signals using random sampling techniques, such that the signal can be reconstructed at the cloud from less number of samples than required by the Nyquist rate [7]. The main benefit of CS for s-Health is providing high compression ratio, while moving the high computational load to the reconstruction phase at the cloud. The second approach, instead, aims at extracting and transmitting the most representative features from the collected data that are associated with the patient's conditions, which substantially decreases the transmitted data size, hence decreasing energy consumption, without affecting the detection of the patient's state [8].

3.2 Event detection at the edge

Given the aforementioned requirements and challenges of s-Health system, this approach aims at enabling energy-efficient delivery of real-time medical data by developing the following:

- A technique for emergency detection at the PDA that identifies the patient's status.
- A selective data transmission strategy that leverages the proposed detection technique to map the acquired data into different transmission modes, considering the patient's status and QoS requirements. This transmission strategy enables transmitting toward the cloud only the essential and representative data, which can further reduce energy consumption in s-Health system.

Data acquisition, feature extraction, and swift classification are the basis of event detection at the PDA. For providing high-intensive monitoring in case of emergency, all collected data from a patient have to be frequently reported to the cloud; whereas, in normal conditions, some critical data features describing the patient's state can be sufficient. Leveraging this fact, it is important to develop a highly accurate classification technique at the PDA, utilizing some features extracted from the gathered data, to provide a reliable detection of the patient's state while requiring low computational complexity.

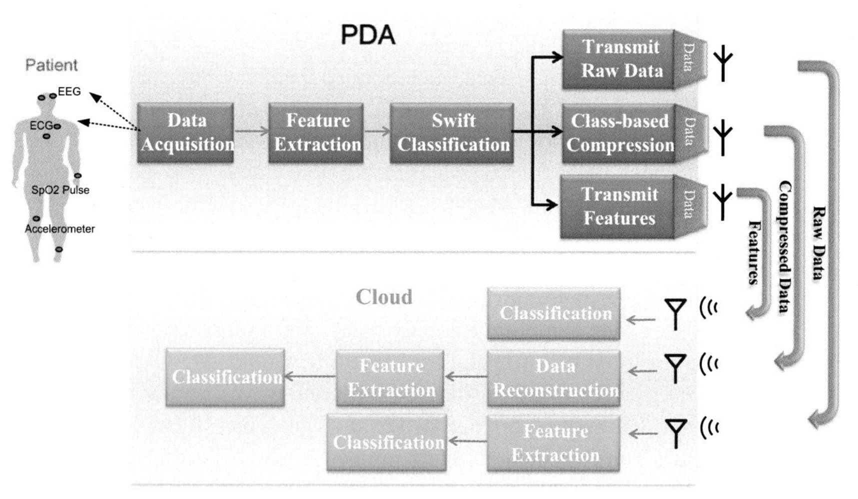

FIGURE 3.3

Energy-efficient data transmission scheme for s-Health system.

Applying this classification to estimate the patient's state at the PDA has two additional advantages. First, it enables a selective data transmission scheme that adopts the most convenient transmission mode according to the detected patient's state (see Fig. 3.3). For instance, if no emergency is detected, the collected data can be further processed to transmit only those features that are essential for patient assessment and treatment. Furthermore, by detecting the state of the patient at the PDA, an efficient class-based compression scheme can be also implemented, which accounts for the data characteristics and the class of the patient to define the best configuration of the compression parameters [9]. Second, a quick emergency notification signal can be sent to notify patient's caregivers in case of emergency.

Decreasing energy consumption due to continuous data transmission and monitoring is the major objective of the proposed s-Health system. Fig. 3.4 assesses the performance of the proposed s-Health system, in terms of PDA's battery lifetime, compared to a mobile-health (m-health) and remote monitoring (RM) systems. In m-health system, the PDA compresses the gathered data, with a fixed compression ratio = 40%, and transfers the processed data to the cloud. In RM system, the PDA is used as a communication hub while conveying all processing tasks to the cloud (i.e., raw data is always sent). In this figure, set of experiments have been conducted considering a practical scenario where a smartphone with full battery is running as a PDA until it runs out of battery. The PDA's power consumption calculations have been estimated using Battery Historian [10]. Moreover, the EEG database in [11] is used, which includes three classes of patients: seizure-free (SF), nonactive (NAC), and active (AC). In our experiments, the compression ratio of s-Health for NAC class is set to 40%. In addition, 10% of the acquired EEG signals belongs to AC class, 20% belongs to NAC class, and 70% belongs to SF class [12]. The selected

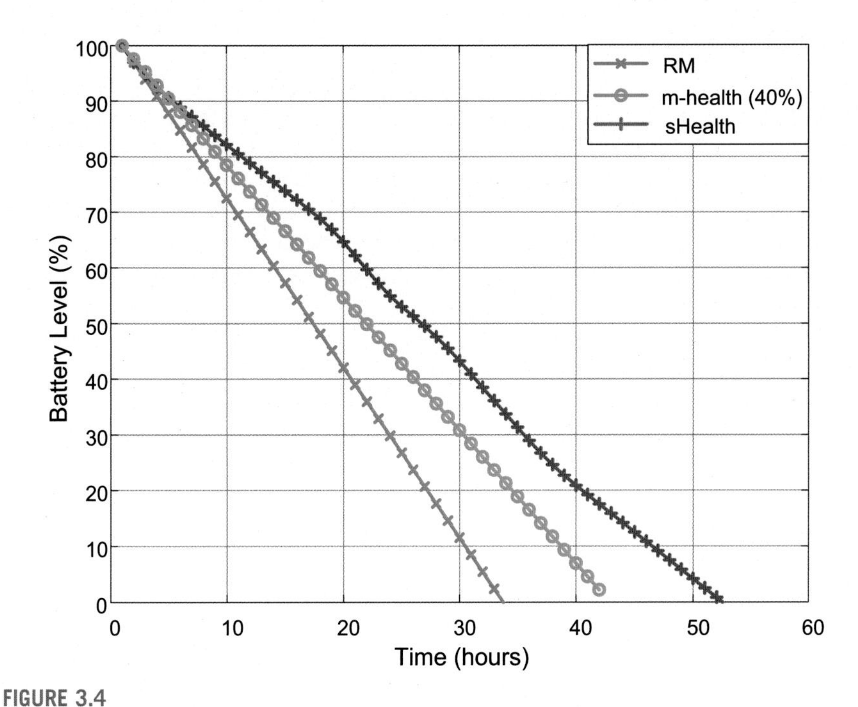

FIGURE 3.4

s-Health, m-Health (with C = 40%), and RM battery lifetimes.

value for the compression ratio has been chosen based on the trade-off between energy consumption and distortion. However, different values can also be selected, taking into consideration the QoS requirements, patient's status, wireless network conditions, and energy budget at the PDA. Fig. 3.4 clearly illustrates that s-Health system provides significant performance improvement in battery lifetime over the RM and m-health methods. For more details about the implemented framework, please refer to [12].

In this context, many machine learning approaches, including supervised, unsupervised and reinforcement learning, were proposed in the literature for the classification of diverse applications. Shortly, supervised learning techniques require two phases: learning from a labeled training dataset, then classifying the testing dataset. Unsupervised learning classifies the acquired datasets into various clusters using the correlation in the input data. The third category is reinforcement learning that leverages real-time learning, which comprises the learning of the environmental conditions and the utilization of the acquired knowledge, to classify the input data [13]. However, some limitations should be considered when applying machine learning techniques in s-Health, including (1) the trade-off between the algorithms' computational complexity and the obtained classification accuracy, (2) the need to process large datasets to maintain high accuracy, (3) it is not trivial to analytically formulate the learned model or to control the learning process.

3.3 RAN-aware optimization

This section discusses the third function through which we can leverage the benefits of MEC, namely, radio access network (RAN)-aware optimization. Thanks to the knowledge on the available RANs quality and user context, the performance of s-Health system can be enhanced by enabling data transfer from edge node to the cloud in an energy-efficient manner, while maintaining a long lifetime of the battery-operated devices. However, this poses several challenges as innovative network association techniques are required, which account for energy efficiency while meeting application's requirements.

A possible approach to tackle the problem of optimizing network association is to adopt a user-centric strategy that enables each user to independently select one or more RANs to use simultaneously. The selection depends on the user's objectives (i.e., energy saving, monetary cost, or service latency), and the characteristics of the available RANs (i.e., throughput, channel quality, and data rate). Furthermore, a dynamic weight update mechanism, as in [14], can be incorporated in the scheme to optimally select the RAN(s) taking into consideration both the user battery level and monetary budget. By doing so, the selection strategy can achieve the desired level of fairness among different user's objectives while significantly enhancing the lifetime of the edge node.

For concreteness, we consider an example of s-Health application where a user has to connect to the available RANs to transfer 10 MB/hour of medical data to the Cloud, and its monetary budget is 45$. Each of the available RANs has different characteristics as follows: RAN1 has a monetary cost per MB $€_1 = 0.3$ \$/MB, and data rate $R_1 = 4$ Mbps; RAN2 has $€_2 = 0.2$ \$/MB, and $R_2 = 3.5$ Mbps, RAN3 has $€_3 = 0$\$/MB, $R_3 = 2.5$ Mbps; RAN4 has $€_4 = 0.1$\$/MB and $R_4 = 3$ Mbps.

Figs. 3.5 and 3.6 show the performance gain of the autonomous selection with weights update (ASWU) algorithm [14] in terms of user lifetime with varying networks association, compared to two baseline algorithms, named ranked network selection (RNS) and autonomous access network selection (AANS). Herein, the user lifetime is defined as the maximum operating time till the mobile user runs out of energy or monetary budget. In RNS, each user computes a score for each of the candidate RANs using its multiobjective function, and the network with the lowest score is selected. In AANS, instead, it considers a multiobjective optimization problem that accounts for user's objectives; however, the weights of different objectives are assumed to be predefined and fixed, while in ASWU a dynamic weights update mechanism is developed to maximize user lifetime. Accordingly, in ASWU and AANS, a PDA can associate to more than one RAN simultaneously instead of being limited to one RAN only (see Fig. 3.5). However, ASWU algorithm efficiently updates the different objectives' weights such that the lifetime is maximized. Hence, as user's monetary budget decreases, the corresponding cost weight increases; a similar behavior is obtained with decreasing energy budget. It follows that ASWU enables the user to dynamically vary its RANs' association to avoid reaching zero energy/money budget. Consequently, Fig. 3.6 illustrates that ASWU can improve the PDA operating time by 15% with respect to AANS, and by 373% with respect to RNS.

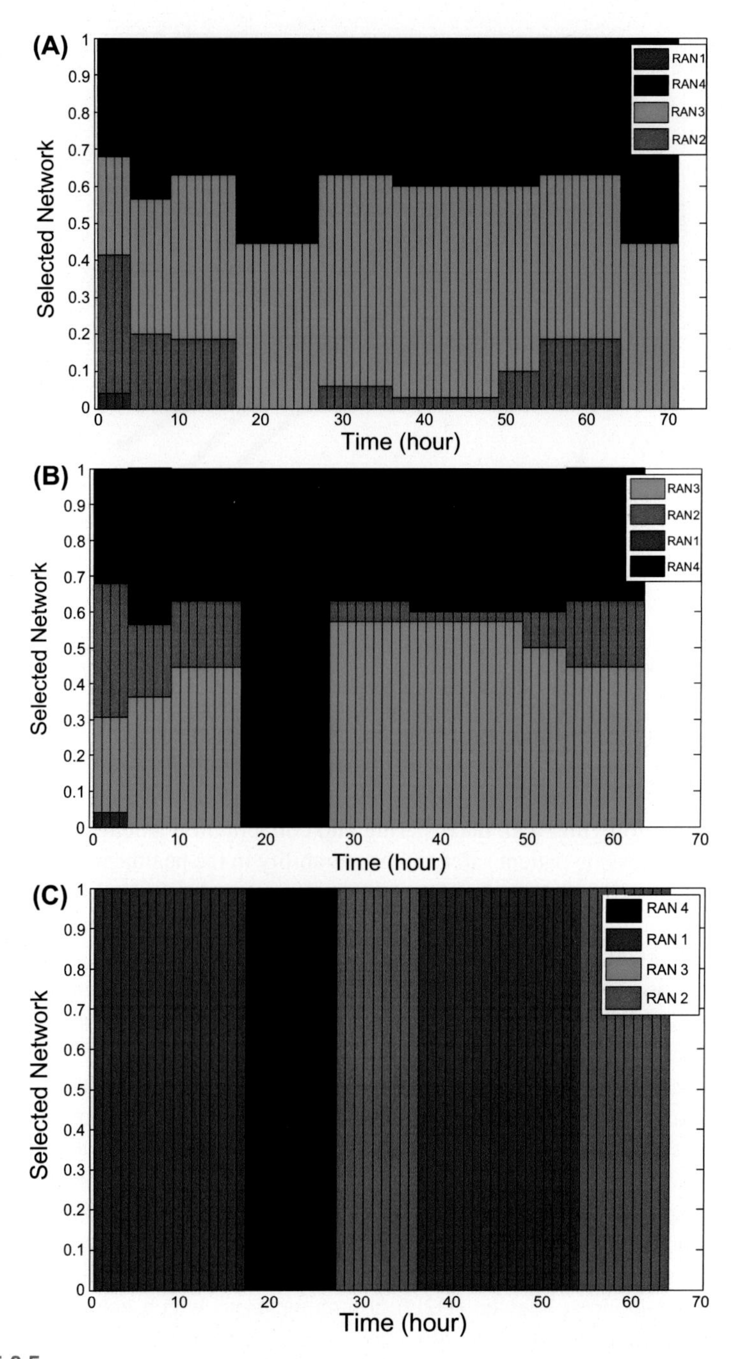

FIGURE 3.5

Selected networks using (A) ASWU, (B) AANS, and (C) RNS.

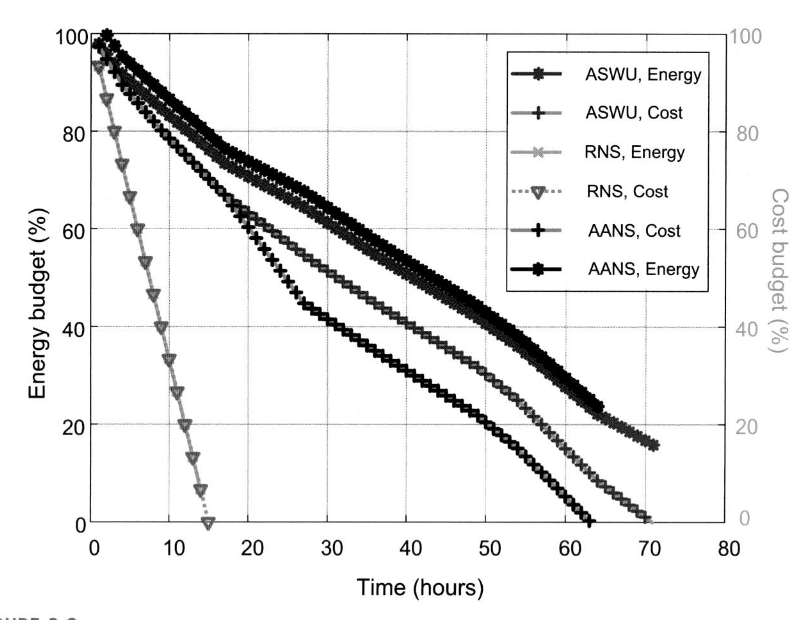

FIGURE 3.6

Energy and monetary budgets assessment for the ASWU, AANS, and RNS schemes.

4. Challenges and open issues

4.1 Cooperative edge

Healthcare is struggling with data sharing and collaboration among different stakeholders for increasing patient safety. Interoperability in the healthcare industry is the ability of diverse health systems to communicate and exchange health-related information (e.g., clinical and administrative data) to provide remote access to a patient's record. For instance, detecting and correlating a patient with a heart attack who has been prescribed medications in different health entities with interacting properties inducing the heart attack, requires coordinated/collaborative data analytic across these entities. However, sharing of the medical data owned by a stakeholder is challenging due to the privacy concerns and the high cost of data transfer.

Accordingly, leveraging cooperative edge that enables the communication between the edges of different stakeholders, which are geographically distributed (such as hospitals, pharmacies, and health institutions), is valuable in threefold. First, it facilitates distributed information management between various stakeholders, thanks to in-network processing at the cooperative edges. Second, it allows the patients to transfer their data toward the cloud with the help of other edge nodes by exploiting device-to-device (D2D) communication, which enhances spectrum and energy efficiency while enabling data transferring in geographically remote areas [15]. Third, it allows a patient's edge to directly communicate with the nearest hospital's edge for getting fast emergency response, without going through the cloud, which also assists in improving monitoring and energy efficiency, as well as operational cost.

4.2 Heterogeneous sources of information

Smart health applications typically rely on data acquisition, aggregation, and real-time analysis of large amount of data from different heterogeneous sources (see Table 3.1). Thus, the proposed s-Health system can be adopted to deal with this challenge through:

- Developing data analytic and vision-based activity recognition techniques at the edge, which support real-time processing of such humongous amount of data to perform knowledge discovery, features selection, clustering, classification, and event detection. This helps also in designing adverse event detection and emergency notification schemes using collected data at the edge to detect patient's status and send a quick emergency notification to notify patient's caregivers or different health entities in case of emergency.
- Designing event-based data transmission strategy that exploits heterogeneous sources of information to provide a compact representation of the relevant data considering not only the intramodality correlation, but also intercorrelation among diverse modalities, QoS requirements, and characteristics of the gathered data. This allows, on one hand, implementing efficient data reduction techniques, on the other hand, reducing the amount of transmitted data, hence saving consumed network bandwidth and transmission energy. We emphasize that deep learning can be a good candidate for such techniques [16], due to its ability to efficiently extract the hierarchical representations of the data and learn the different order features from heterogeneous sources of information.
- Leveraging computational intelligence at the edge for implementing data fusion algorithms (including probabilistic methods, artificial intelligence, and theory of belief) for emergency detection and patient tracking. This multimodal fusion can significantly enhance the overall system reliability through detecting several distress situations.

Several studies have been presented in the field of behavioral signal processing and recognition methodologies for inference of complex human behavior and psychological states, leveraging multimodal data [17], in particularly audio-visual and physiological sensing data. In [18], authors present a case study on chronic pain measurement and management exploiting various sensing modalities including: activity monitoring from accelerometer and location sensing, audio analysis of speech, and image processing for facial expressions. However, many challenges are still open when we come to the s-Health. First, it is not straightforward to consider multiple active and passive modalities in s-Health system, where energy consumption is a limiting factor. Second, noise artifacts emanate from internal sources, such as muscle activities, or from external sources, such as interference and signals offset, have severe impact on data quality [19].

Table 3.1 Common data types in healthcare applications.

Data type	Examples
Physiological	Electroencephalogram (EEG), electrocardiogram (ECG), blood pressure, electromyography (EMG), electrooculography (EOG), blood oxygen, respiratory rate, temperature.
Healthcare information	Smoking, gene sequence, family history, protein sequence, diabetes, medical image.
Behavioral	Sleep time, frequency of wake up, walking speed, rest time and frequency, eating time.
Environmental	Surveillance video, pollution density, weather conditions, noise level.

5. Conclusion

This chapter proposed our vision of a s-Health system that leverages multiaccess edge computing. The proposed system architecture can significantly promote the system performance through efficiently handling the massive amount of data generated by different medical/nonmedical devices at the network edge. While addressing the large data size and constrained energy availability of such devices, we also account for both applications and data characteristics. In particular, in-network processing like compression and event detection has shown great effect on reducing the amount of data transmitted to the cloud, hence addressing one of the main bottlenecks in s-Health system. In this context, this chapter proposed some effective approaches and computing tasks to be implemented at the edge for optimizing energy consumption, emergency response time, and bandwidth utilization. Finally, it highlighted the main challenges and opportunities of applying edge computing within s-Health that are worth to further investigated.

Acknowledgments

This work was made possible by Qatar University Grant QUHI-CENG-19/20-1. The findings achieved herein are solely the responsibility of the authors.

References

[1] Kay M, Santos J, Takane M. mHealth: new horizons for health through mobile technologies. World Health Organization; 2011. p. 66–71.

[2] Solanas A, et al. Smart health: a context-aware health paradigm within smart cities. IEEE Communications Magazine August 2014;52(8):74–81.

[3] Shi W, Cao J, Zhang Q, Li Y, Xu L. Edge computing: vision and challenges. IEEE Internet of Things Journal Oct 2016;3(5):637–46.

[4] Multi-access edge computing," http://www.etsi.org/technologies-clusters/technologies/multi-access-edge-computing, July 2017.

[5] Yazicioglu R, Torfs T, Merken P, Penders J, Leonov V, Puers R, Gyselinckx B, van Hoof C. Ultra-low-power biopotential interfaces and their applications in wearable and implantable systems. Microelectronics Journal 2009:1313–21.

[6] Awad A, Mohamed A, Chiasserini C-F, Elfouly T. Distributed in-network processing and resource optimization over mobile-health systems. Journal of Network and Computer Applications 2017;82:65–76.

[7] Chiang J, Ward RK. Energy-efficient data reduction techniques for wireless seizure detection systems. Sensors 2014;14(2):2036–51.

[8] Awad A, Saad A, Jaoua A, Mohamed A, Chiasserini CF. In-network data reduction approach based on smart sensing. In: IEEE global communications conference (GLOBECOM); December 2016. p. 1–7.

[9] Abdellatif AA, Mohamed A, Chiasserini C. Automated class-based compression for real-time epileptic seizure detection. In: 2018 wireless telecommunications symposium (WTS); April 2018. p. 1–6.

[10] Analyzing power use with battery historian. Available from: https://developer.android.com/topic/performance/power/battery-historian.html, April 2018.

[11] Andrzejak R, Lehnertz K, Rieke C, Mormann F, David P, Elger C. Indications of nonlinear deterministic and finite dimensional structures in time series of brain electrical activity: dependence on recording region and brain state. Physical Review E 2001; 64:061907.

[12] Abdellatif AA, Emam A, Chiasserini C, Mohamed A, Jaoua A, Ward R. Edge-based compression and classification for smart healthcare systems: concept, implementation and evaluation. Expert Systems with Applications 2019;117:1–14.

[13] Alsheikh MA, Lin S, Niyato D, Tan HP. Machine learning in wireless sensor networks: algorithms, strategies, and applications. IEEE Communications Surveys Tutorials Fourthquarter 2014;16(4):1996–2018.

[14] Awad A, Mohamed A, Chiasserini CF. Dynamic network selection in heterogeneous wireless networks: a user-centric scheme for improved delivery. IEEE Consumer Electronics Magazine January 2017;6(1):53–60.

[15] Awad A, Mohamed A, Chiasserini CF, Elfouly T. Network association with dynamic pricing over D2D-enabled heterogeneous networks,. In: IEEE wireless communications and networking conference (WCNC); March 2017. p. 1–6.

[16] said AB, Al-Sa'D MF, Tlili M, Abdellatif AA, Mohamed A, Elfouly T, Harras K, O'Connor MD. A deep learning approach for vital signs compression and energy efficient delivery in mhealth systems. IEEE Access 2018;6. 33 727–733 739.

[17] Narayanan S, Georgiou PG. Behavioral signal processing: deriving human behavioral informatics from speech and language. Proceedings of the IEEE 2013;101(5):1203–33.

[18] Aung MSH, Alquaddoomi F, Hsieh CK, Rabbi M, Yang L, Pollak JP, Estrin D, Choudhury T. Leveraging multi-modal sensing for mobile health: a case review in chronic pain. IEEE Journal of Selected Topics in Signal Processing 2016;10(5): 962–74.

[19] Sweeney KT, Ward TE, McLoone SF. Artifact removal in physiological signals: practices and possibilities. IEEE Transactions on Information Technology in Biomedicine 2012;16(3):488–500.

CHAPTER

Energy-efficient EEG monitoring systems for wireless epileptic seizure detection

4

Ramy Hussein[1], Rabab Ward[2]

[1]*Postdoctoral Fellow, Electrical and Computer Engineering, The University of British Columbia, Vancouver, BC, Canada;* [2]*Professor Emeritus, Electrical and Computer Engineering, The University of British Columbia, Vancouver, BC, Canada*

1. Introduction

Wireless electroencephalogram (EEG) monitoring systems have been used for remote seizure detection applications. These systems capture, process, and transmit the EEG data wirelessly to a server, where the data are analyzed and epileptic seizures, if any, are detected [1]. Several wireless EEG-based seizure detection methods have been developed and shown to achieve promising seizure detection performance [2–6]. The success of such methods holds promises for better epilepsy management. For example, when a patient is experiencing a seizure, these systems would alert the family members and/or caregivers so that they could take care of him/her during and after the seizure attack. They may clear the physical space around the patient of any sharp or hard objects to prevent injury, turn the patient on his/her side to ensure that his/her airway remains clear during the seizure, and time the seizure and call for medical help if any complications arise [7].

For the wireless seizure detection systems to have a clinical value in epilepsy monitoring and management, it is crucial to develop a reliable method for processing and transmitting ambulatory EEG signals. With recent advances in wireless and electronics technologies, ergonomic, lightweight, and comfortable designs of wireless EEG headbands have become an increasingly viable alternative to the traditional wired devices for EEG monitoring. A wireless EEG sensor unit is a miniaturized, battery-powered device that captures, processes, and transmits EEG signals wirelessly to a server at the receiver side, where the data are stored and further data analysis is carried out. A major limitation of such wireless EEG devices is their limited battery's lifetime. For a typical EEG montage with 24 EEG electrodes and using a sampling rate of 400 Hz and a 16-bit analog-to-digital converter, a data rate of 150 kbps is generated. Given such a data rate, the traditional way of transmitting the entire raw EEG data to the server side is not feasible. This is because wireless transmission is extremely hungry in terms of power consumption. As shown

Energy Efficiency of Medical Devices and Healthcare Applications. https://doi.org/10.1016/B978-0-12-819045-6.00004-2

in Ref. [8], wireless EEG data transmission accounts for ~70% of the total power consumption of wireless EEG devices.

To improve the power consumption in wireless EEG data transmission, the size of the data to be transmitted should be significantly reduced. A possible technique is to apply data compression to the raw EEG signals before their transmission. This can be done by deploying compressive sensing or wavelet transform compression techniques [9,10]. Dynamic EEG channel selection, as an alternative data reduction method, has also been used for wireless seizure detection applications [11,12]. Another recent data reduction approach is to perform feature extraction on the sensor unit and only transmit the EEG features that are pertinent to epileptic seizures [6,13]. However, the previously mentioned data reduction methods encounter two main challenges. First, some of these methods are computationally expensive and the energy they consume to process the EEG signals at the sensor side becomes comparable to that needed for their wireless transmission. Second and most importantly, the reduction in the amount of transmitted data may result in a nontrivial loss in the EEG information content, which would negatively affect the seizure detection performance.

To address these challenges, this chapter describes a computationally simple, energy-efficient, and pseudolossless EEG data reduction method for wireless seizure detection applications. This method is denoted by missing at random-expectation maximization or MAR-EM [14]. This method consumes less energy than the state-of-the-art methods while achieving a superior seizure detection performance. At the sensor side, the size of data is reduced by deleting some data points selected randomly. This process is computationally simple and effectively reduces the power consumption in data transmission, and hence elongate the battery life. At the server side, a machine learning algorithm, named expectation maximization, is employed to recover the missing (randomly deleted) data points. This method can be applied in any of the three different types of EEG monitoring frameworks. These frameworks are based on one of the following procedures: (1) the transmission of all raw EEG data, (2) the transmission of the compressed EEG data, or (3) the transmission of extracted EEG features. We show that, for each of these frameworks, this method can considerably reduce the total power consumption of the wireless EEG devices while maintaining high seizure detection performance. A reduction of ~60% in power consumption can be achieved while obtaining a seizure detection accuracy of 95%–99%.

2. Research background

Only a few studies have been conducted on energy-efficient EEG monitoring systems for wireless seizure detection applications [15]. In Ref. [15], Nia et al. developed an energy-efficient scheme for a personal health monitoring system. It uses a set of different biomedical sensors such as accelerometer, blood pressure, heart rate, and also EEG. The proposed scheme incorporated compressive sensing, sample aggregation, and anomaly-driven transmission to reduce the power consumption of wireless EEG transmission. Experimental results demonstrated that the proposed scheme can achieve considerable energy savings by simply accumulating the sensor data before transmitting them to the server side.

In Ref. [11], Shih et al. proposed an EEG channel (sensor) selection method for energy-efficient ambulatory seizure monitoring applications. They used a machine learning algorithm to automatically identify the EEG channels that included relevant information for seizure detection. Using a clinical EEG dataset taken from 16 patients, the proposed algorithm was examined and its performance was compared to an earlier study. Results demonstrated that the proposed algorithm can effectively reduce the number of EEG channels from 18 to 6 while maintaining comparable seizure detection accuracy (97%). The average detection latency, however, increased from 7.8 seconds to 11.2 seconds. In a similar study [12], Faul et al. presented a dynamic EEG channel selection to reduce the overall power consumption in wireless seizure detection systems without compromising the detection accuracy. Different combinations of EEG channels were tested and the combination that achieved the best seizure detection rate was selected for further analysis. Experimental results showed that the proposed channel selection method achieves power savings up to 47% without affecting the seizure detection performance.

In Ref. [13], Chiang et al. introduced energy-efficient data reduction methods for reducing transmission data in a wireless EEG seizure detection system. They studied two data reduction methods: *compressive sensing* and *on-board feature extraction*. The achieved performance was assessed in terms of seizure detection accuracy and total power consumption (the trade-offs between the detection accuracy and power consumption were also discussed). The results demonstrated that by transmitting only the EEG features that are pertinent to seizure patterns, the power consumed by wireless transmission could be significantly reduced. This helped extend the battery lifetime of the sensor node by a factor of 14 while maintaining the same seizure detection performance as the traditional method (with a seizure detection sensitivity of 95%).

In Ref. [6], Hussein et al. developed on-board data reduction approach that extracted low complexity and high-level application-based EEG features at the sensor side. Specifically, the EEG spectrum was segmented into five frequency subbands; numerous combinations of these subbands were selected as feature vectors, and the EEG classification was carried out using k-nearest neighbor. Simulations have revealed that α and δ EEG rhythms formed the most representative feature vector needed for accurate detection of epileptic seizures. Satisfactory seizure detection accuracy of 92.47% classification accuracy was obtained. Moreover, the proposed approach in Ref. [6] is found to outperform conventional data streaming and compression methods in terms of total power consumption and seizure detection performance.

3. EEG feature extraction and classification

3.1 EEG data and subjects

The MAR-EM method described later was tested on the EEG dataset provided by Bonn University [16]. In this chapter, we address the three-class classification problem for distinguishing the following EEG categories: *Normal EEG* is recorded from five healthy subjects who do not suffer from epilepsy, *Interictal EEG* is recorded from five epileptic patients during seizure-free intervals, and *Ictal EEG* is recorded from five patients experiencing active seizures. Each EEG class has 100 single-channel EEG signals, each of 23.6-seconds duration. All the EEG signals have been filtered, amplified, and sampled at 173.6 Hz and digitized using a 12-bit analog-to-digital converter (ADC).

3.2 EEG feature extraction

Feature extraction is a dimensionality reduction process that removes redundancy in raw data to facilitate the subsequent analysis and classification processes, and in some cases lead to better human interpretations. Efficient feature extraction methods should thus be developed to eliminate the data ambiguity and decrease the computation cost of the classification problem. In this chapter, we use a conventional frequency-domain-based feature extraction method to examine the effectiveness of the MAR-EM method. The deployed feature extraction method has low computational complexity and results in distinguishable EEG features that rightly characterize the original data.

The captured EEG signals are first transformed to the frequency domain using fast Fourier transform (FFT) and then typical features are computed from the EEG rhythms Delta (δ), Theta (θ), Alpha (α), Beta1 (β_1), Beta2 (β_2), and Gamma (γ). The frequency ranges of these rhythms are 0.5–4, 4–8, 8–13, 13–22, 22–35, and >35Hz, respectively. The features of mean, median, minimum, maximum, standard deviation, skewness, kurtosis, and average power are computed from each EEG rhythm and then combined together to form the feature vector.

Fig. 4.1 shows the frequency spectrum of noisy and clean EEG signals. The noisy signal is corrupted with synthetic muscle artifacts, and its signal-to-noise ratio is 0 dB. The figure clearly shows that the spectrum of the noise-free EEG signals is localized in the low-frequency range of 0–50 Hz while the spectrum of the noisy EEG signals is disseminated over the entire frequency range. This motivated us to use the EEG features extracted from the EEG rhythms below 50 Hz.

Our experiments have shown that the features extracted from the EEG rhythms δ, θ, and $\beta 1$ only result in a representative feature vector needed for accurate detection of epileptic seizures.

3.3 EEG classification

As clarified in Section 3.1, the classification problem is to distinguish between normal (nonseizure) EEG, interictal (between seizures) EEG, and ictal (during a seizure) EEG signals. Given that each EEG class has 100 signals, a total of 300

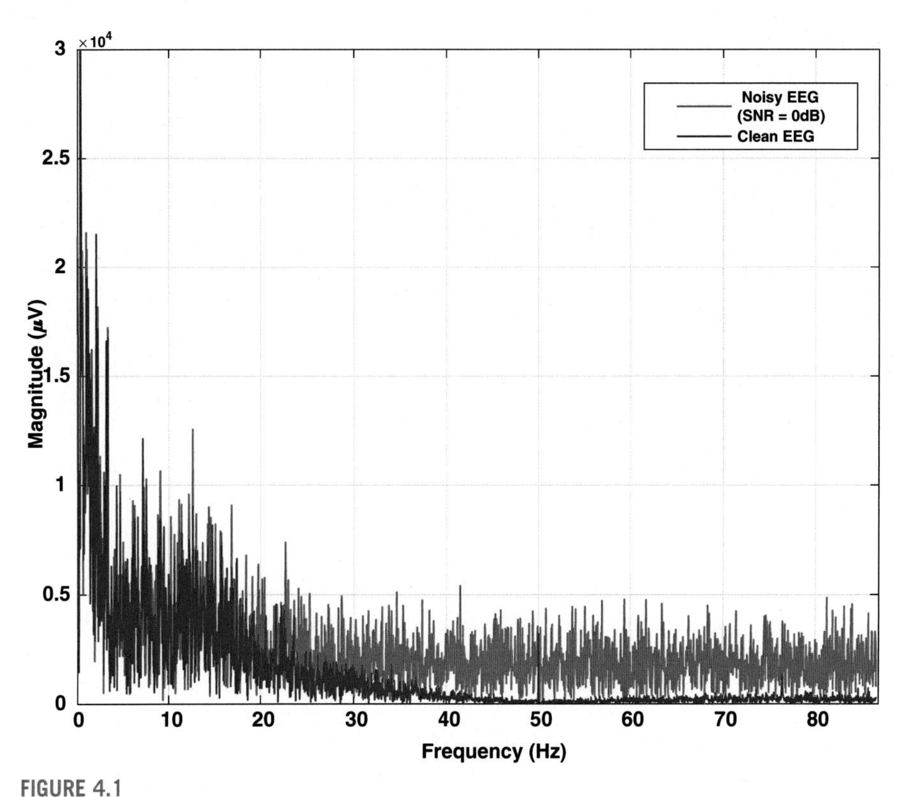

FIGURE 4.1

Frequency spectra of noisy and clean EEG signals.

EEG signals were used for training and testing the proposed seizure detection method. For every EEG signal, feature extraction was carried out and the selected features were concatenated to construct a 24-element feature vector.

To determine whether a newly observed feature vector is representative of normal, interictal, or ictal activity, a multiclass classification model was used. We examined the performance of several classifiers and found that random forest (RF) achieves the superior seizure detection performance. The RF integrates a set of tree predictors, each has its own weight and is considered as an individual classifier [17]. The overall classification accuracy is computed based on the classification outputs of all trees. Principally, the correct class is identified based on the vote of most trees. In this study, an RF classifier with 10 trees was used in the classification of all feature types.

The classification performance of the proposed wireless seizure detection systems was evaluated on a per-subject basis using leave-one-out cross-validation [18], which works as follows: In each round, the feature vectors from all, but one, of the subjects' data were used as the training set to train the classifier. The feature vectors from the withheld data of the subject were then used to test the classifier.

This process was repeated until the data of the tested subjects were withheld once. This test reflects the ability of the seizure detection method to generalize from the training set and to classify new unobserved data.

The seizure detection performance was assessed in terms of the following:

- **Sensitivity**, which measures the proportion of actual positives that are correctly identified as such (e.g., the percentage of seizure epochs that are correctly classified as seizure epochs by the classifier).
- **Specificity**, which measures the proportion of actual negatives that are correctly identified as such (e.g., the percentage of nonseizure epochs that are correctly classified as nonseizure epochs by the classifier).
- **Classification accuracy**, which measures the number of correct predictions made divided by the total number of predictions made, multiplied by 100 to turn it into a percentage.

4. Proposed energy-efficient method

This section describes how to encode the data at the sensor side so that the total power consumption is reduced to $\sim$60%. It also demonstrates how to decode the data at the server side to retrieve the original missing data.

4.1 Missing at random—sensor side

To increase the sensors' lifetime of a wireless EEG monitoring system, their power consumption should be significantly reduced. Thus, we modified the existing EEG sensor units to include a new module named "missing at random (MAR)" [19]. This module deletes some EEG data points randomly, and thereby reduces the size of the EEG data that need to be transmitted to the data server (where the data are recovered and seizure detection is carried out). This data reduction method helps to decrease the power consumption in wireless transmission without increasing the processing power consumed at the EEG sensor nodes.

Fig. 4.2 depicts an example of resulting patterns of missing values for raw EEG data, where the missing data points (variables) occur at random locations. The observed and missing data points are indicated by white and red colors, respectively. The vertical axis of Fig. 4.2 corresponds to 100 EEG epochs, while the horizontal axis corresponds to the number of data points in each EEG epoch. For convenience, only 0.3 seconds (50 EEG time instances) are plotted. In this figure, each EEG epoch is missing 30% of its data points. that is, 15 out of 50 EEG data points are missing.

4.2 Expectation maximization—server side

To classify the data at the server side, the missing values should be estimated first. Many statistical methods have been proposed to interpolate incomplete data. Expectation maximization (EM) is one of the most promising algorithms that can

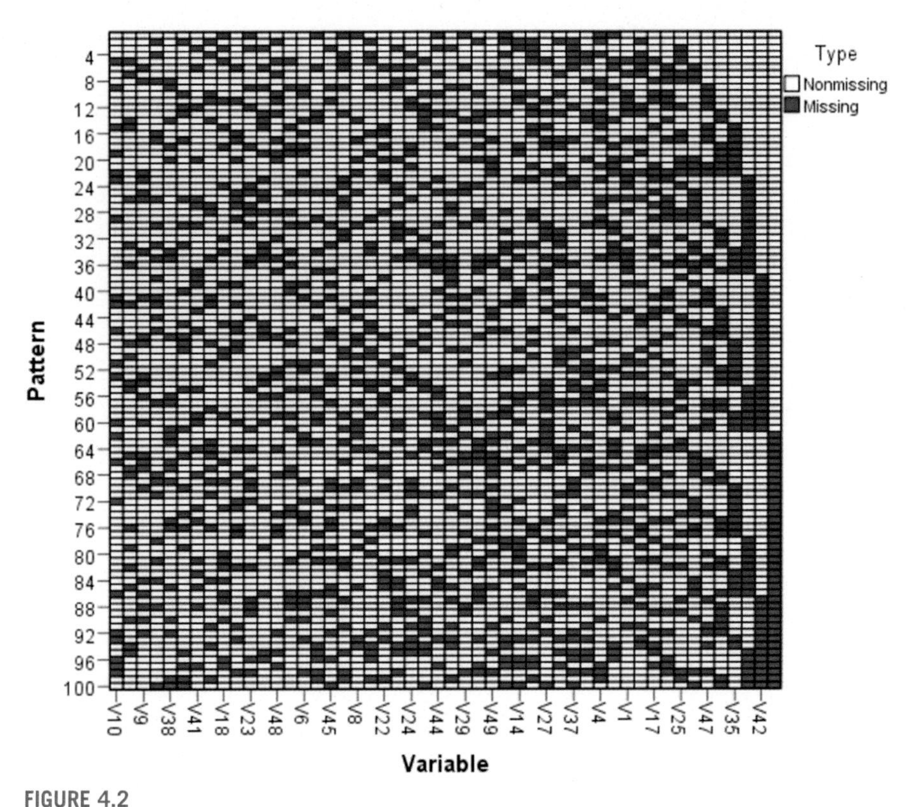

FIGURE 4.2

Missing value patterns of 100 EEG epochs.

effectively estimate missing entries [20]. Given a statistical model, EM is utilized to estimate the missing data Z given an observed data X. This can be formulated as that of finding the model parameter ξ such that the conditional probability $\mathcal{P}(X|\xi)$ is a maximum. $\mathcal{P}(X|\xi)$ can be represented in terms of the missing values z_i as follows:

$$\mathcal{P}(X|\xi) = \sum_{z_i} \mathcal{P}(X, z_i|\xi) \qquad 4.1$$

where ξ denotes the parameters of the probabilistic model we try to find.

EM finds the maximum likelihood estimate of the previously mentioned marginal likelihood by iteratively applying the following two steps:

1. Expectation step (E-step): Estimate $Q(\xi|\xi^t)$—the conditional expectation of the log-likelihood based on the current estimate of the model parameters ξ^t:

$$Q(\xi|\xi^t) = \mathbb{E}_{z|X,\xi^t}\left[\log\left(\sum_{z_i} \mathcal{P}(X, z_i|\xi)\right)\right] \qquad 4.2$$

2. Maximization step (M-step): Calculate the parameter that yields the maximum log-likelihood estimate with respect to ξ:

$$\xi^{t+1} = \arg\max_{\xi} Q(\xi|\xi^{t}) \quad 4.3$$

The EM method implemented in IBM SPSS software is used to recover the missing EEG data points. The performance of the EM algorithm is examined at different proportions of missing data starting by 10% and ending by 50%.

5. EEG data transmission in a wireless seizure detection system

The scientific literature reports on three main frameworks of wireless EEG monitoring systems. The first framework captures the raw EEG data and sends it as is to the server side. The second framework uses a compression method to reduce the data size and then transmits the compressed data to the server side (where data reconstruction and analysis are performed). The third framework applies feature extraction on the sensor side, but only the selected features are then transmitted to the server side.

In this section, we show how to modify the three previously mentioned frameworks so that their overall power consumption at the sensor side is reduced, while maintaining high seizure detection accuracy at the server side. Fig. 4.3 depicts the

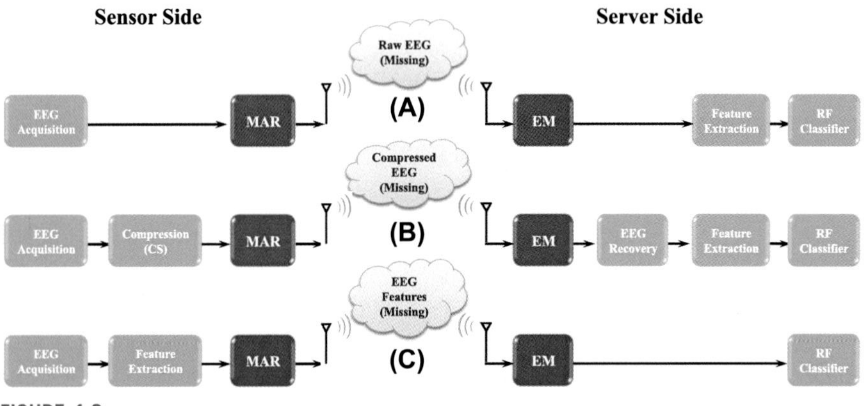

FIGURE 4.3

Proposed energy-efficient EEG monitoring methods for wireless epileptic seizure detection. (A) MAR is applied to the entire raw EEG data before transmission, (B) MAR is applied to the compressed EEG data before transmission, and (C) MAR is applied to the distinctive EEG rhythms before transmission. EM is used at the server side to retrieve the missing values.

modified frameworks. The sensor side is amended to include the MAR module, which deletes some EEG data points at random. On the server side, the EM algorithm is used to estimate the values of missing EEG data points. To evaluate the effectiveness of the MAR-EM scheme, the detection accuracy and total power consumption are evaluated for the three main frameworks.

5.1 Transmission of entire raw EEG data

The traditional framework of wireless EEG monitoring systems transmits the entire raw data (RD) to the server side. The main drawback of this system lies in the massive power consumption used by the wireless transmission. Fig. 4.3 top branch depicts the modified framework (by adding an MAR module at the sensor side and an EM module at the receiver side) to maintain high seizure detection performance with much less power consumption. The total power consumption of the RD framework, denoted by P_{tot}^{RD}, is computed as follows:

$$P_{tot}^{RD} = P_E^{RD} + P_T^{RD} \tag{4.4}$$

where P_E^{RD} denotes the power consumed by encoding (acquiring the data and amplifying it) and P_T^{RD} is the power required for data transmission. For a multichannel EEG monitoring system, P_E^{RD} is given by

$$P_E^{RD} = C(P_{Amp} + P_{ADC}) \tag{4.5}$$

where C is the total number of EEG channels (electrodes), P_{Amp} and P_{ADC} correspond to the power consumption of the amplifier and the analog-to-digital converter, respectively.

In addition, the power required for data transmission, P_T^{RD}, is given by

$$P_T^{RD} = C(f_s RJ) \tag{4.6}$$

where f_s is the sampling rate, R is the *ADC* resolution, and J is the transmission energy per bit.

After modifying the *RD* framework to include the *MAR* module, the power needed for transmission is scaled by $\frac{L}{N}$factor to be:

$$P_T^{RD} = C\left(\frac{L}{N} f_s RJ\right) \tag{4.7}$$

where L and N are the lengths of the missing and original data, respectively. From Eqs. (4.5) and (4.7), the total power consumption of the *RD* framework is then given by

$$P_{tot}^{RD} = C\left(P_{Amp} + P_{ADC} + \frac{L}{N} f_s RJ\right) \tag{4.8}$$

5.2 Transmission of compressed EEG data

EEG compression techniques are utilized to reduce the data size and hence the transmission power. Compressive sensing (CS) provides the most effective techniques that have been used for EEG data compression [21]. The compression rate (CR) is computed as N:M, where N is the length of the original data and M is the length of the compressed data produced by CS. A compression rate of 5:1 is used here so that the transmission power is significantly reduced to the fifth of its original value. At the server side, a CS reconstruction method is used to recover the original signal. Fig. 4.3 middle branch elaborates on the modified CS-based EEG encoding framework after adding the MAR and EM modules. The overall power consumption of such a compressed data (CD)-based EEG monitoring framework, denoted by P_{tot}^{CD}, is given as follows:

$$P_{tot}^{CD} = P_E^{CD} + P_T^{CD} \tag{4.9}$$

where P_E^{CD} and P_T^{CD} correspond to the encoding and transmission power of the CD-based framework.

As compressing the data using compressive sensing necessitates deploying the random number generator (RNG) and matrix multiplication (MM) modules, the encoding power $P_E^{CD}$is computed as

$$P_E^{CD} = C(P_{Amp} + P_{ADC}) + P_{RNG} + P_{MM} \tag{4.10}$$

where P_{RNG} and P_{MM} denote the power consumption of the RNG and MM blocks, respectively.

The transmission power is also derived as

$$P_T^{CD} = C\left(\frac{1}{CR}\frac{L}{M}f_s R\ J\right) \tag{4.11}$$

where CR denotes the compression rate and equals $\frac{N}{M}$.

By taking the sum of Eqs. (4.10) and (4.11), the total power consumption of the modified CD-based EEG monitoring framework is obtained as

$$P_{tot}^{CD} = C\left(P_{Amp} + P_{ADC} + \frac{1}{CR}\frac{L}{M}f_s RJ\right) + P_{RNG} + P_{MM} \tag{4.12}$$

5.3 Transmission of EEG features

Recently, the studies in Refs. [6,13] have presented novel EEG encoding schemes. They applied feature extraction to the EEG data at the sensor side. Once the EEG data is captured, the features relevant to seizures are extracted and sent to the server side (where EEG classification is performed). The proposed schemes yield considerable power savings in both data encoding and transmission. They, however, showed limited seizure detection performance. Here, we modify these on-sensor processing schemes to incorporate the MAR and EM modules, with the ultimate objective of saving more power and achieving better seizure detection accuracy. MAR is applied to the captured EEG features to further reduce the transmitted data size and certainly the transmission power. At the server level, EM is used to estimate the missing EEG features. Fig. 4.3 bottom branch describes the modified on-sensor feature extraction framework.

The feature extraction method described in Section 3.2 is implemented at the sensor level. The frequency spectrum of EEG data is first obtained, and the EEG rhythms of δ, θ and β_1 are then transmitted to the server side. This is where the features that are pertinent to seizures are computed from these rhythms and used as inputs to the RF classifier.

Thus, the encoding power consumption of the EEG features (EEGF)-based encoding framework is expressed as [6]:

$$P_E^{EEGF} = C\left(P_{Amp} + P_{ADC} + \frac{3N}{2} S \log_2(N)\right) \tag{4.13}$$

where $\frac{3N}{2}\log_2(N)$ is the computational complexity incurred by FFT, and S is the net power required to perform one FFT instruction.

The transmission power, denoted by P_T^{EEGF} is also computed as

$$P_T^{EEGF} = C\left(\frac{F}{N}\frac{L}{F} f_s RJ\right) \tag{4.14}$$

where F represents the number of the frequency components in the EEG rhythms δ, θ, and $\beta 1$ (i.e., 390).

Thus, the overall system power consumption of EEGF-based EEG monitoring framework is represented as

$$P_{tot}^{EEGF} = C\left(P_{Amp} + P_{ADC} + \frac{3N}{2} S \log_2(N) + \frac{L}{N} f_s RJ\right) \tag{4.15}$$

6. Power consumption evaluation and seizure detection performance

To evaluate the effectiveness of the MAR-EM energy-efficient method, the total system power consumption is computed for each framework individually. The standard performance metrics of sensitivity, specificity, and classification accuracy are also evaluated at the server side. Matlab is used to quantify the power consumption values at the sensor side, and the WEKA software is used to assess the seizure detection performance at the server side.

6.1 Raw EEG data streaming

For the raw EEG streaming framework, the wireless EEG sensor unit is in charge of data acquisition and transmission to the server side, where diagnosis and detection processes are conducted. As shown in Fig. 4.3 top branch, we modify the traditional

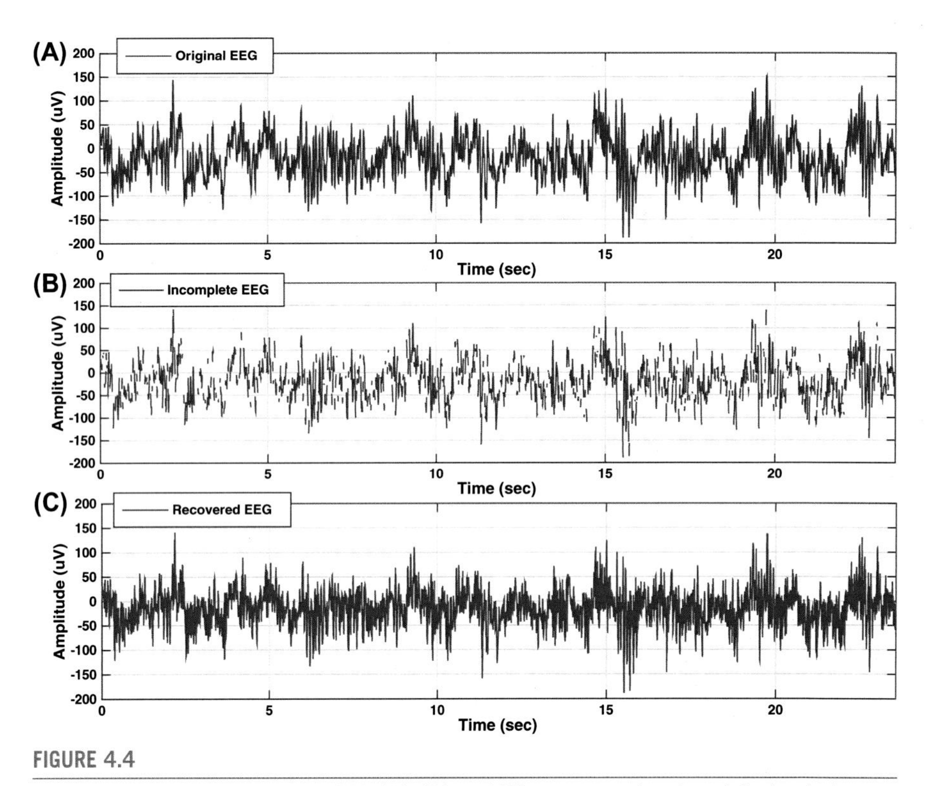

FIGURE 4.4

Recovery of raw EEG data using EM: (A), (B) and (C) correspond to the original, missing, and recovered raw EEG data, respectively.

model of raw EEG streaming by integrating the MAR and EM modules. The MAR module intentionally deletes some data points and hence lessens the size of the EEG data that need to be transmitted to the data server. The EM module is employed at the server side to recover the missing values. Fig. 4.4A shows a raw EEG signal of 23.6-second duration. The impact of the MAR module is depicted in Fig. 4.4B, where the EEG signal is missing 30% of its samples. Fig. 4.4C demonstrates the efficacy of the EM method to recover the original EEG signal. The root mean square error between the original and recovered EEG signals is found to be 0.012, which verifies that both signals are similar.

To prove the power saving capability of the modified raw EEG streaming framework, the overall system power consumption is evaluated and compared to those of the state-of-the-art methods. The values of the system parameters C, P_{Amp}, P_{ADC}, R, and J are 24, 2.9 μW, 0.2 μW, 12 bit/sample, and 50 nJ/bit, respectively [22,23]. The standard seizure detection performance metrics are also evaluated and listed together with the power consumption values in Table 4.1. The first two rows display the sensitivity, specificity, classification accuracy, and total power consumption achieved by the state-of-the-art methods presented in Refs. [6,13]. Despite the high seizure

Table 4.1 Seizure detection performance and total power consumption of raw data (RD) model.

Method	Sensitivity (%)	Specificity (%)	Accuracy (%)	$P_{tot}^{RD}(mW)$
Chiang et al. (2014) [13]	94.91	99.83	97.37	32.50
Hussein et al. (2015) [6]	95.14	90.06	93.54	32.50
Proposed, MAR (00%)	99.50	99.50	99.50	32.50
Proposed, MAR (10%)	97.25	99.78	99.18	22.20
Proposed, MAR (20%)	96.84	98.88	98.26	19.80
Proposed, MAR (30%)	94.95	99.76	97.54	17.40
Proposed, MAR (40%)	91.18	98.00	96.80	15.00
Proposed, MAR (50%)	89.66	96.78	95.15	12.70

detection accuracy achieved by Chiang's method [13], the EEG sensor unit is found to consume a considerable total power of 32.50 mW. Hussein's method achieves a comparable power consumption with lower seizure detection performance [6].

The MAR-EM, on the other side, can effectively achieve the optimal trade-off between the power consumption of the wireless EEG sensor nodes and the seizure detection performance at the server side. It consumes less power while ensuring a comparable seizure detection accuracy. The utilized MAR module can reduce the size of the raw EEG data by different percentages. The larger the percentage of the deleted data points, the more power saving the MAR-EM method achieves. MAR (XX%) indicates that a percentage of XX% is missing from the data. The detection accuracy and power consumption are evaluated at different missing percentages starting with 00% (no missing points) and ending by 50%. It is clearly observed that the MAR-EM method can significantly reduce the total power consumption without compromising the epileptic seizure detection performance. We recommend the use of the MAR-EM method with a 30% missingness. It achieves a comparable seizure detection accuracy to Chiang's method [13], but with ~50% savings in the total power consumption.

6.2 Compressed EEG data streaming

Given the proposed CS-based EEG monitoring system shown in Fig. 4.3B and the power consumption models derived in Eq. (4.10), (11), and (12), the total estimated

Table 4.2 Seizure detection performance and total power consumption of compressed data (CD) model.

Method	Sensitivity (%)	Specificity (%)	Accuracy (%)	P_{tot}^{CD}(mW)
Chiang et al. (2014) [13]	91.82	99.40	95.61	7.46
Hussein et al. (2015) [6]	81.37	87.55	85.95	7.46
Proposed, MAR (00%)	95.32	99.00	97.46	7.46
Proposed, MAR (10%)	92.75	98.27	96.08	6.41
Proposed, MAR (20%)	92.14	97.66	94.92	5.35
Proposed, MAR (30%)	92.14	97.34	94.92	4.30
Proposed, MAR (40%)	89.20	89.20	89.19	3.26
Proposed, MAR (50%)	84.90	84.90	84.88	2.21

power consumption values are shown in Table 4.2. The power consumption of the random number generator (P_{RNG}) and matrix multiplication (P_{MM}) are 3 μWμ W and 352 μW, respectively [24]. The overall system detection performance and power consumption are evaluated against different missing percentages in the compressed data. The numerical results reported in Table 4.2 verify the effectiveness of the MAR-EM-based seizure detection method over the state-of-the-art methods. A significant decrease in the data transmission cost is achieved without seriously affecting the detection accuracy. For example, the fourth row of Table 4.2 demonstrates that the MAR-EM method with 30% missing data can achieve a detection accuracy comparable to Chiang's method [13] while reducing the power consumption from 7.46 mW to 4.30 mW.

6.3 EEG features streaming

The MAR-EM-based energy-efficient EEG monitoring framework shown in Fig. 4.3C gives superior results over the state-of-the-art on-sensor systems. The system parameter *S* takes the value of 66 pW [25]. Before integrating the MAR module (i.e., 00%MAR), the obtained classification accuracy is 99.38% and the overall power consumption is 2.30 mW. The use of the MAR module can further reduce the number of frequency components that should be transmitted to the server side, and hence yields a sufficient reduction in total power consumption. The quantitative results listed in Table 4.3 explain the decrease in the detection accuracy along the

Table 4.3 Seizure detection performance and total power consumption of EEG features (EEGF) model.

Method	Sensitivity (%)	Specificity (%)	Accuracy (%)	P_{tot}^{EEGF}
Chiang et al. (2014) [13]	94.91	99.83	97.37	2.30
Hussein et al. (2015) [6]	94.82	90.06	92.47	2.30
Proposed, MAR (00%)	100.00	98.75	99.38	2.30
Proposed, MAR (10%)	99.40	99.40	99.33	1.84
Proposed, MAR (20%)	99.27	98.24	98.62	1.32
Proposed, MAR (30%)	96.88	98.97	98.26	0.90
Proposed, MAR (40%)	95.98	98.56	97.54	0.58
Proposed, MAR (50%)	93.55	97.43	96.90	0.25

corresponding missing percentages. It can be seen that, for an MAR of 40% missing points, the MAR-EM method can fulfill a better seizure detection accuracy than achieved by Chiang's method [13] while consuming ~25% only of its power consumption.

7. Limitations and recommendations

The proposed energy-efficient wireless seizure detection systems incorporate the expectation maximization algorithm for data recovery at the server side. This algorithm works efficiently when there is only a small percentage of missing data and the dimensionality of the data is not too big [26]. For data with high dimensionality and/or high missing percentage, the EM algorithm runs very slow. This is because the expectation step is computationally expensive, and it converges extremely slowly as the procedure approaches a local maximum [27]. We also noticed that the reconstruction accuracy of the EM algorithm is considerably reduced when more than 50% of the data is missing. The resultant high reconstruction error has a serious impact on the extracted EEG features, which negatively affects seizure detection accuracy. We thereby recommend using the energy-efficient MAR-EM systems for offline EEG monitoring applications only. We also advise keeping the missing data percentage below 50% to ensure small reconstruction error and fast convergence.

8. Summary and conclusion

In this chapter, an energy-efficient method for wireless epileptic seizure monitoring and detection was described. The Missing at Random (MAR) algorithm was used at the sensor side to reduce the energy cost of EEG data transmission. At the server side, we utilized the Expectation Maximization (EM) method to reconstruct the missing values and accurately recover the missing data. Further, we used an efficient feature extraction method to select distinguishable EEG features for epileptic seizure detection. The EEG rhythms δ,θ and β_1 were selected to compute a set of descriptive statistical features that represent the input to the Random Forest classifier. The superiority of the proposed MAR-EM scheme over the existing baseline methods is verified. It achieves 42–75% lower power consumption while maintaining comparable seizure detection accuracy as state-of-the-art methods.

References

[1] Casson AJ, Yates DC, Smith SJ, Duncan JS, Rodriguez-Villegas E. Wearable electroencephalography. IEEE Engineering in Medicine and Biology Magazine 2010;29(3):44–56.

[2] Young CP, Liang SF, Chang DW, Liao YC, Shaw FZ, Hsieh CH. A portable wireless online closed-loop seizure controller in freely moving rats. IEEE Transactions on Instrumentation and Measurement 2010;60(2):513–21.

[3] Ludvig N, Medveczky G, Kuzniecky R, Illes G, Devinsky O. System and device for seizure detection. U.S. Patent No. 7,885,706. Washington, DC: U.S. Patent and Trademark Office; 2011.

[4] Borujeny GT, Yazdi M, Keshavarz-Haddad A, Borujeny AR. Detection of epileptic seizure using wireless sensor networks. Journal of Medical Signals and Sensors 2013;3(2):63.

[5] Sawan M, Salam MT, Le Lan J, Kassab A, Glinas S, Vannasing P, Nguyen DK. Wireless recording systems: from noninvasive EEG-NIRS to invasive EEG devices. IEEE Transactions on Biomedical Circuits and Systems 2013;7(2):186–95.

[6] Hussein R, Mohamed A, Alghoniemy M. Energy-efficient on-board processing technique for wireless epileptic seizure detection systems. In: International conference on computing, networking and communications (ICNC). IEEE; 2015. p. 1116–21.

[7] Stacey WC, Litt B. Technology insight: neuroengineering and epilepsydesigning devices for seizure control. Nature Reviews Neurology 2008;4(4):190.

[8] Yazicioglu RF, Torfs T, Merken P, Penders J, Leonov V, Puers R, Van Hoof C. Ultralow-power biopotential interfaces and their applications in wearable and implantable systems. Microelectronics Journal 2009;40(9):1313–21.

[9] Casson AJ, Rodriguez-Villegas E. Toward online data reduction for portable electroencephalography systems in epilepsy. IEEE Transactions on Biomedical Engineering 2009;56(12):2816–25.

[10] Daou H, Labeau F. Dynamic dictionary for combined EEG compression and seizure detection. IEEE Journal of Biomedical and Health Informatics 2013;18(1):247–56.

[11] Shih EI, Shoeb AH, Guttag JV. Sensor selection for energy-efficient ambulatory medical monitoring. In: Proceedings of the 7th international conference on mobile systems, applications, and services. ACM; 2009. p. 347–58.

[12] Faul S, Marnane W. Dynamic, location-based channel selection for power consumption reduction in EEG analysis. Computer Methods and Programs in Biomedicine 2012;108(3):1206–15.

[13] Chiang J, Ward R. Energy-efficient data reduction techniques for wireless seizure detection systems. Sensors 2014;14(2):2036–51.
[14] Hussein R, Ward R, Wand ZJ, Mohamed A. Energy efficient EEG monitoring system for wireless epileptic seizure detection. In: 15th IEEE international conference on machine learning and applications (ICMLA). IEEE; 2016. p. 294–9.
[15] Nia AM, Mozaffari-Kermani M, Sur-Kolay S, Raghunathan A, Jha NK. Energy-efficient long-term continuous personal health monitoring. IEEE Transactions on Multi-Scale Computing Systems 2015;1(2):85–98.
[16] Andrzejak RG, Lehnertz K, Mormann F, Rieke C, David P, Elger CE. Indications of nonlinear deterministic and finite-dimensional structures in time series of brain electrical activity: dependence on recording region and brain state. Physical Review E 2001; 64(6):061907.
[17] Breiman L. Random forests. Machine Learning 2001;45(1):5–32.
[18] Kearns M, Ron D. Algorithmic stability and sanity-check bounds for leave-one-out cross-validation. Neural Computation 1999;11(6):1427–53.
[19] Little RJ, Rubin DB. Statistical analysis with missing data, vol. 793. John Wiley & Sons; 2019.
[20] Bilmes JA. A gentle tutorial of the EM algorithm and its application to parameter estimation for Gaussian mixture and hidden Markov models. International Computer Science Institute 1998;4(510):126.
[21] Donoho DL. Compressed sensing. IEEE Transactions on Information Theory 2006; 52(4):1289–306.
[22] Murmann B. A/D converter trends: power dissipation, scaling and digitally assisted architectures. In: IEEE custom integrated circuits conference. IEEE; 2008. p. 105–12.
[23] Semiconductor F. XS110 UWB solution for media-rich wireless applications. 2004. http://www,_freescale,_com/files/microcontrollers/doc/factsheet/MOTUWBFS,_pdf.
[24] Holleman J, Otis B, Bridges S, Mitros A, Diorio C. A 2.92 W hardware random number generator. In: Proceedings of the 32nd European solid-state circuits conference. IEEE; 2006. p. 134–7.
[25] Hussein R, Mohamed A, Alghoniemy M, Awad A. Design and analysis of an adaptive compressive sensing architecture for epileptic seizure detection. In: 2013 4th annual international conference on energy aware computing systems and applications (ICEAC). IEEE; 2013. p. 141–6.
[26] Gupta MR, Chen Y. Theory and use of the EM algorithm. Foundations and Trends in Signal Processing 2011;4(3):223–96.
[27] Moon TK. The expectation-maximization algorithm. IEEE Signal Processing Magazine 1996;13(6):47–60.

CHAPTER

5 Intelligent energy-aware decision-making at the edge in healthcare using fog infrastructure

R. Leena Sri, PhD [1], V. Divya, BE, MTech [2]

[1]Thiagarajar College of Engineering, Madurai, Tamilnadu, India; [2]Reseach Scholar, CSE, Thiagarajar College of Engineering, Madurai, Tamilnadu, India

1. Introduction

In the advent of today's evolution on Industry 4.0, there has been a surge in the number of intelligent devices that in-turn has raised the amount of data produced. Almost all present-day devices are smart and capable of communication and computation. With the increasing amount of data, highly capable devices are needed to meet the needs of the industries. Although the capability of data handling could have increased, the utmost need is to provide decisions in real time. This need of the hour has led to the emergence of various computing paradigms, each with advantages in its own way. When used for the right application, the computing platform can be advantageous in providing accurate decisions and increasing the overall efficiency of the application.

The evolution of computing paradigms started from the traditional enterprise computing, which was a centralized computing framework that lagged in performance in terms of both energy and time. The platform was highly inflexible and was slow in development. Thus, the "fat" client/server model evolved giving the advantage of flexibility, but still was not cost-effective and the configuration cost was also high. This led to the development of the "thin" client/server model, which still had the disadvantage of the centralized facility and the hindrance to resource sharing. The evolutions of the various paradigms are given in Fig. 5.1, which shows the roadmap to the present-day technology that is being used for real-time data processing.

The need for resource utilization and the raise in the distributed computing paradigms led to the introduction of grid computing. This computing platform consists of connected devices, which are a geographically remote system that forms a single network. This computing facility was said to be a virtual supercomputer as it combines the power of the systems that form the network. In a similar way came the era of cluster computing that forms a connected network usually a LAN, and takes advantage of the parallel computing. This has further improved the performance

Energy Efficiency of Medical Devices and Healthcare Applications. https://doi.org/10.1016/B978-0-12-819045-6.00005-4

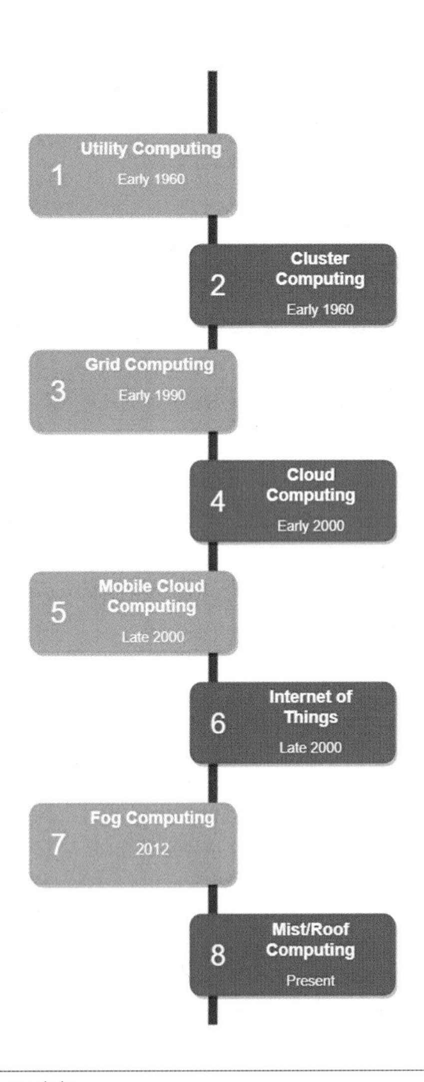

FIGURE 5.1

Evolution of computing models.

of the hosted application by providing high processing power, fault tolerance, and availability of resources to the application [1].

Taking on the advantage of the distributed computing and resource-sharing methods, the provisioning model called the utility computing rose to take over the computing world. This model helps in the maximization of the resource utility at a minimal cost. As the name suggests, the model provisions services on demand to the users and the charge is based on the utility of the resources. With the technology improvement in terms of computational power, bandwidth utility, and the network speed, the utility computing evolved to provide software as a service, which later emerged to cloud computing.

Cloud computing became the choice of platform for almost all applications because it provides the advantage of provisioning virtualized resources on demand to the users. Almost all applications such as banking, weather forecasting, and space use the cloud platform to access the resources that involve the platform, infrastructure, and any type of resource needed for the application.

Alongside the development of the cloud platform, which almost became everything as a service, the development of the Internet of Things (IoT) has also led to the current intelligent applications. The IoT became an interconnected system of almost everything with a unique identification and the ability to transfer data in the network. The extensive use of IoT applications has led to the development of numerous sensors for a variety of applications. Thus, in each application, as the number of sensors or devices increases, the data also increase and these data were not always analyzed, and the big data analytics plays a major role in every real-world application.

The data analytics and the decisions taken have to be provided in near-real time for the efficiency of the application and to make complete use of the data. The streaming data collected from the sensors have their expiry time bound in most of the applications after which the collection is of no use to be analyzed in real time. Once the next slot of data approaches, the previous data expire, and thus, the decision-making has to be in real time for better performance of the system.

This has again led to the new paradigm of computing called the edge/fog computing that prefers decision-making at the edge devices [2]. The fog layer lies on-premise at a one-hop distance from the sensors. This infrastructure removes the hindrance of sending data from the sensors to a remote location, which increases the latency and the network utility of the application. The major features of fog computing that help in the improvement of the performance of the application are given in Fig. 5.2.

With the increase in the computational and communication capabilities, the most important factor that has to be considered is the energy efficiency of the node. The end devices that are said to be the edge nodes and the fog nodes are nodes with limited capabilities and have to be efficient enough to conduct the computations.

The healthcare application is a crucial application that has to be both latency-aware and energy-aware for better performance of the application. Our work proposed here is of two layers, the one that takes care of the computation right at the edge device and the upper layer takes care of the intermediate intelligence and the data filtering to ensure efficient bandwidth utilization. The work that is being offloaded at the fog layer has to be intelligent enough such that the energy of the nodes is also taken into account. The load has to be balanced in an intelligent fashion such that the resource availability is maintained throughout. Thus, the aspect of energy-aware computing is very essential in the real-time scenario.

The remainder of the chapter is organized as follows. Section 2 deals with the related works, which consist of the relation between IoT and cloud, computing and research aspects in the use of this paradigm, and the introduction to the new

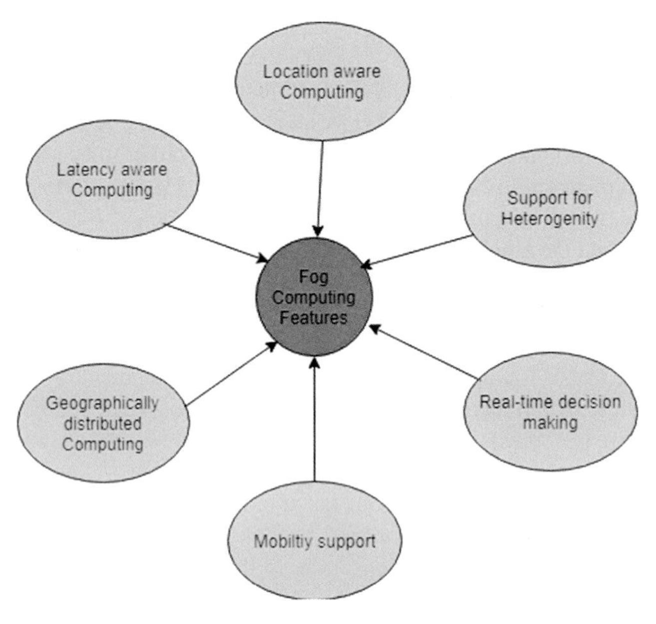

FIGURE 5.2

Fog computing features.

paradigm of fog computing. The section presents the architecture perspective, which describes the proposed framework built to handle the data for healthcare applications. It also involves the energy parameters that are to be considered in the work. The section deals with the algorithmic perspective of the work for energy-aware computing. The initial section consists of the literature that deals with the use of the heuristics to handle the energy efficiency, which is then compared with the proposed reinforcement learning (RL)-based optimization in the later section. Finally, the chapter ends with the conclusion and the scope for further research on the topic.

2. Related works

There has been development in the era of IoT application from the late 2000 and in the present days, and the prominence of use has increased in almost all real-time applications. Almost all objects of everyday life from buildings to vehicles are capable of generating data and communicate with each other and sometimes even capable of computations when in need. These data are very useful and have to be collected and analyzed in real time for further intelligence on the application. During the initial days, the applications and the decision-making were considered as big data applications and decision-making depended on platforms such as the Hadoop to improve the efficiency of the application. With the increase in the advancements in cloud computing, the IoT data were handled by a cloud server at a remote location.

The IoT applications generate huge streaming data that are time bound, and the huge data are to be stored and processed, which is rendered by the cloud server. The major advantage of the collaboration of the cloud and the IoT devices is the ability to access data from a remote location and to process them on demand and also scale the application resources on demand at an affordable cost. In the case of industrial IoT applications, the use of the remote cloud server helps in the allocation of resources in various geographical areas and efficient monitoring of the same. The industries require minimal investment in the infrastructure setup and on-demand scaling of their applications.

With the introduction of automation to the health industry, there have been applications of monitoring and remote healthcare system with cloud as the backbone infrastructure. The work by Ref. [3] brings out the advantage of IoT and cloud collaboration in healthcare with patient monitoring as the application. The work deals with the management and decisions of the data from the mobile and wearable sensors, where the data had been pushed to the cloud and the decision is sent back to the concerned authority. Although the IoT-cloud infrastructure may seem to be highly efficient, the major challenges in terms of real-life applications especially in the health sector by the authors [4] include as described.

- The migration of the data between servers to support the mobility of the devices is both costly and time consuming.
- Delayed real-time communication between the IoT device and the cloud server
- Disrupted interoperability between the services and infrastructures
- Difficulty in the maintenance of accountability of resources in case of mobility and remote computations

With these persisting issues, the major need in most of the IoT applications is the need for an immediate decision on receiving sensitive data in an intelligent way. Thus, a new computing paradigm was introduced to handle sensitive applications and data near the data source. Cisco has come up with a solution to handle this challenge termed as "Fog Computing." This computing technique allows decision-making and part of the computing near the edge and pushes only relevant data for further intelligence to the cloud [5].

The fog nodes are the computing devices that can be deployed anywhere near the data source for analysis. The nodes may include controllers, switch, routers, embedded servers, and video surveillance cameras. The nodes must be capable of storage, computation, and network connectivity. Table 5.1 gives the comparison of the capabilities of the fog and cloud infrastructures.

2.1 Role of fog computing in IoT

The features of the Internet of Things can be used as an advantage to the IoT applications. The location awareness, low latent computations near the edge can be an important feature in applications like augmented reality, real-time surveillance, and gaming. In addition to this, the advantage of distributed computing and the

Table 5.1 Computing paradigm comparison.

	Fog infrastructure	Cloud infrastructure
Architecture	Decentralized	Mostly centralized
Computing power	Limited/low	High
Response time	Milliseconds/seconds/ minutes	Minutes/days/weeks based on complexity
Data storage	Transient	Permanent
Geographic area coverage	Limited	Wide spread
Data security	High	Limited
Application latency	Very low	Considerably high for latency sensitive application

The advantages of the fog architectures are taken up by the proposed framework.

mobility feature of fog computing can be used in major applications such as traffic monitoring. The data can be processed near the edge and can be transferred in need to the neighboring nodes with acceptable latency and increased performance.

As the fog nodes are heterogeneous, they can be deployed in a distributed environment and the data can be collected to a common pool and the immediate processing can be done at the edge devices. With these significant features of the fog nodes and fog computing, the recent application that has been under the limelight of research is the connected vehicles. The fog nodes are being used in various aspects such as the traffic analysis, location-aware actions, and low latency decision-making and geographically distributed data analytics for a complete application.

In a similar way, the smart traffic management system helps in the internetworking of the sensor data and the computations in near real time. The traffic regulation is based on approaching pedestrians and vehicles in real time at the edge. The feature of mobility helps in enhancing the interoperability of the fog nodes that are geographically distributed. The use of the edge computation also adds an advantage of improving the bandwidth utility by reducing the data being sent to the upper layer, which is the cloud in a remote place. After the immediate decision has been rendered, the data from each of the fog node/cluster are sent to the remote cloud server for further intelligence and proactive functioning of the application for selected functionality.

The working of the application does not depend on the fog infrastructure only. The combination of the fog and the cloud infrastructure helps in the improvement of the performance of the application. The fog layer assists in providing latency-sensitive decisions and the cloud aids in the further intelligence of the application needed for long-term improvement in functionalities. The time span of decision-making also depends on the geographical distribution of the nodes and the type of analytics needed. For instance, the real-time analytics takes time of about seconds to minutes and the transactional analytics may need longer time of hours or days.

Thus, the deployed fog nodes are heterogeneous in nature and have the ability to handle a variety of intelligence. In all applications, the dataflow is bidirectional and the nodes have to be capable of both communication and computation.

3. Edge/fog computing—architectural perspective

To realize any computing paradigm, there has to be a valid architecture on which the applications are hosted. The performance of the application is dependent on the underlying architecture as equally as the algorithm that drives the application. There have been various studies in building up a fog architecture to enhance the overall performance of the system.

3.1 Building of fog architecture

The basic architecture of the fog architecture is given in Fig. 5.3. The lower layer of the architecture includes the IoT layer, which is the edge devices that may or may not be intelligent. The intermediate fog layer aids in the reduction of involvement of the cloud and sending the data to a remote location. There have been several studies that had implemented the fog infrastructure for their application to enhance the performance in terms of latency, energy utilization, and bandwidth utility.

There have been several works to perfect the architecture for efficient service rendering and to take the complete advantage over the fog layer. The initial work

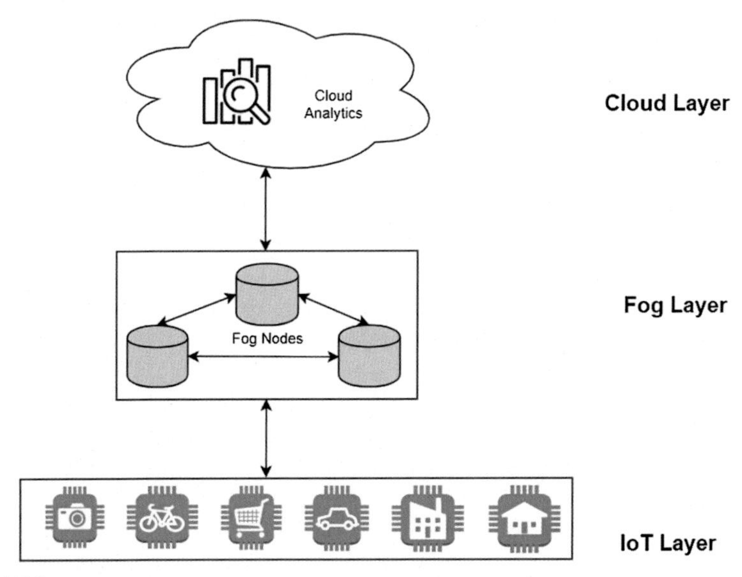

FIGURE 5.3

Basic fog architecture.

on proposing a high-level architecture started with [6], where an intermediate layer of fog nodes comprising the routers and networking devices was used for task offloading. The IoT layer comprises smartphones and connected vehicles. The fog nodes were capable of delivering on-demand compute instances based on the requests. The work tries to enhance traffic monitoring by enhancing the performance even in case of mobility.

Later, the research progressed as done by Ref. [7] to extend the platform as a service architecture (PaaS) to be supported for the fog architecture. The architecture consisted of a cloud controller with a separate execution engine that is a part of the fog layer. The applications running were containerized to increase the ability of scalability and mobility. The notable contribution to the work includes resource provisioning by a layer-based architecture, which extends the existing PaaS and enables the interaction between the multiple layers with the help of REST protocol. The controller was used for the purpose of efficient orchestration. This architecture was used to simulate the application of fire detection and provide decisions in near real time.

As the real-time applications increased, the need for mobility of the nodes became a greater need. The work by Ref. [8] has been the head start in research to provide mobility among fog nodes and migration of resources. In this work, the fog nodes comprised the cloud-hosted virtual machines (VMs) for computation. The pitfall of this work is that, even though the data are made available in case of the mobility of the application, the migration of the VMs causes the communication delay in the application. This work has clearly explained the need for the collaboration of the cloud and the fog layer where the cloud takes cares care of the compute-intensive tasks and the fog takes care of the latency-sensitive tasks.

As the data from the IoT layer increase, there is an utmost need for proper distribution of the load among the nodes. The work by Ref. [9] involves the basic three-layer architecture with the IoT, fog, and cloud layers. The request from the bottom layer is first sent to the fog layer and when the resources cannot be provisioned by the fog layer, the request is forwarded to the cloud layer. The proposed work deals with the use of fog servers wherein a server manager helps in the management and allocation of VMs. The advantage of the work deals with the low latency resource provisioning by access to the nearest fog server on the reception of a request.

As the research progresses, another dimension to the research was added for the efficiency of the system in an energy perspective. The fog nodes are usually low-end devices with limited energy availability in terms of power, memory, and CPU utilization. The work by Ref. [10] deals with the energy-efficient interoperability of the smart devices in the fog layer. The architecture in this work consists of the raspberry pi as a gateway that helps in the communication of the fog and cloud layer.

Further, the work by Ref. [11] has introduced the use of software-defined networking (SDN) for a global view of network architecture. This architecture has been shown to be used in the creation of wide-area or personal networks. These basics have been used in the creation of our proposed testbed.

In continuation, the research by Ref. [12] deals with the context-aware node placement to enhance the efficiency of the application. The work deals with the dissipation of load, such that the energy and the latency of the application are reduced. The simulation of the work is done for real-time problems from smart cities. The placement of the virtual devices consists of lightweight containers that help in the reduction of both the latency of the application and the energy consumption of the devices. As the deployment is in the form of lightweight containers, the fog nodes with minimal capacity shall handle the computations effectively with the least possible use of energy.

3.2 Proposed architecture

Our proposed work consists of a real-time testbed building to realize the fog architecture. The networking component of the architecture is taken care of by the Cisco Nexus switch that is the SDN component in the architecture. This provides a global view of the network and helps in maintaining an overall eagle-eye perspective of the architecture. The fog nodes are realized using the built VLANs, which act as the intermediary decision-making layer between the edge layer and the cloud layer. The architecture also consists of the distributed controllers that help in fault tolerance and seamless load balancing of the application. The proposed architecture is shown in Fig. 5.4. The attached monitoring node helps in the traffic and data collection that helps in providing intelligence to the application.

Now that the architecture has been built, the application focused on in our work is on the healthcare sector. There have been various research works in the field of healthcare. The need for near real-time decision-making and rendering of decisions

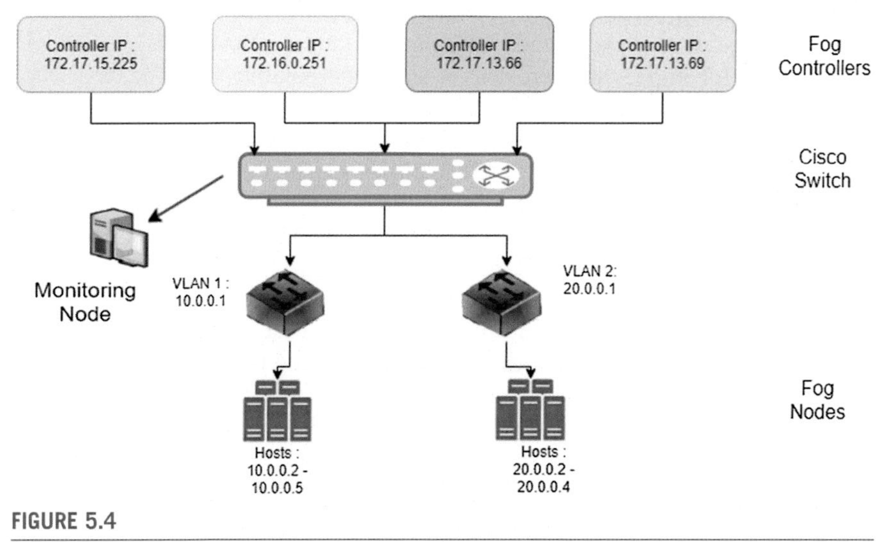

FIGURE 5.4

Proposed fog architecture.

had made edge computing platform suitable for healthcare applications. The following section shows the related literature in the healthcare sector and the advantage of edge computing for the proposed application.

3.3 Edge computing in the healthcare sector

As computation and communication power increase, so do the applications that are in need of low latency processing. As the number of IoT devices and the need for portability increase especially in health sectors, moving from the traditional cloud implementations to edge computing becomes essential. Healthcare has become partly dependent on the IoT devices for daily monitoring of the patients and to keep track of the patient records. The data collected on the patients are sensitive and need to be private. The privacy of the data is ensured by edge computing, such that the data are not pushed to a third-party cloud and the data are not exposed. The data from the IoT devices in this case, the wearable devices are interfaced with the edge data center, which is often built as an extension of the existing network and helps in providing low latency decision-making even in cases of low or poor connectivity.

The other dimension of edge computing in IoT is the need for data filtering, which helps in the conservation of the bandwidth. The data collected from the IoT devices are streaming data at varying velocity, depending on the application or the period of collection. Not all data collected are anomalies, and sending huge data from the devices to the cloud increases the bandwidth utility of the application. Thus, the fog nodes are presented with the ability of data filtering, which detects the anomalies in the data and helps push only the filtered data to the cloud for further analytics. The analytics, when carried in the remote cloud, leads to an increase in the communication latency of the application, which results in a late response to the changes in sensitive parameters of the patient. Proper offloading of applications to the edge or fog devices shall prove to be a potential solution to reduce the latency and the bandwidth utility of the application.

In this regard, there have been several studies to improve the quality of service (QoS) and the quality of experience (QoE) of the users in the healthcare sector and improve the performance and the cost-effectiveness of the applications. The survey by Ref. [13] shows that the applications of edge computing have been creating a great impact in the field of healthcare.

As the integration of the IoT and the edge platform keep increasing, the components needed for the implementations increase. There has been a constant increase in the areas of edge software and analytics by top platforms such as AWS, Cisco, and apache. There have also been explorations toward increasing and standardizing the fog implementations initiated by platforms such as Open Fog Consortium and Cloud Foundry. The integration of the IoT and the cloud platforms were also taken up by initiatives such as Google IoT Cloud, Azure IoT, and IBM Watson. The need for storage of the data also increased with the introduction of the databases such as the HaperDB, MongoDB, and eXtremeDB, which hold importance in the integration of the data with the cloud. With these increases in the supportive platforms, the deployment of the edge servers, fog nodes, etc. has become effective at an affordable cost.

4. Optimization—an algorithmic perspective

The major concern in today's applications is the need for energy conservation in any application hosted. The energy may denote power consumption, optimal resource utility that in-turn shall help in reducing the power consumption, etc. based on the hosted application. The following section elaborates on the need for energy-aware computing on the edge and the various algorithmic perspective to reduce the energy consumption of the hosted application.

4.1 Energy-aware perspective of edge computing

The initial work on the edge platform started with the efficient task allocation to the nodes to help improve their efficiency in terms of latency and throughput of tasks handled by the nodes. The authors in Ref. [14] have formulated this process as an NP-hard problem by their comparison with the bin-packing problem and have tried formulating an efficient task scheduling on the edge nodes.

As the data are being handled in the new perspective at the edge, selected tasks are offloaded to the edge devices to be handled with the least possible latency. As the load to the edge devices increases, the performance of the application also has to be taken care of and the deterioration can happen in case of high-energy depletion. This may even lead to the unavailability of resources due to energy-depleted nodes. Thus, the notion of energy plays an important role in the edge platform for reliability and the improved performance of the hosted application.

The energy of the nodes has to be taken care of in case of both resource allocation and task offloading. There has been research to improve energy efficiency during resource allocations; the work by Ref. [15] resolves the issue of energy-efficient allocation with the help of Markov decision process for service migration and task offloading.

In most cases, the algorithm tends to be computationally intensive, thereby degrading the performance in case of dynamic task allocation and when the load is received in busts. Thus, the authors [16] have tried to take advantage of the local search algorithms to overcome the pitfalls in cases of dynamic allocation. In continuation, there have been uses of the various genetic algorithms, such as ant colony optimization and particle swarm optimization, which have proved to be better than the previous studies in terms of effective task allocation [17,18].

In addition, the major task of filtering happens at the edge layer, which sends only the image of interest to the upper layer that helps in improving the energy efficiency of the edge devices by improvement in bandwidth utility and minimal data transfer to the upper layer.

4.2 Meta-heuristic methodologies

The research on the energy-saving in almost all real-time applications has been looked upon as a scheduling problem. In most of the applications, the factors under consideration are computation time, resource utility, and energy efficiency. The problem is often formulated as an NP-Hard problem and for time-efficient decision-making, the literature suggests the use of heuristic algorithms for better efficiency of the hosted application. The following sections revolve against the various meta-heuristic algorithms used in the literature for improving energy efficiency.

4.2.1 Particle swarm optimization-based energy optimization

Energy demand has been on the rise in almost all applications in our day-to-day life. This optimization of energy, in turn, leads to the overall expenditure reduction, thereby increasing the performance of the application in terms of cost and QoS and in the current sense, QoE. One of the major concerns in the present-day smart environment is the check on the energy utilization and the heat dissipation from the automated systems in almost all the equipment that we use. As the hardware gets intelligent, the devices are embedded with the computing and communication capabilities, which lead to the heat dissipation and the high-energy utilization. As most of the devices used today are battery operated, the energy resources are limited and these devices are responsible to handle the large streaming data. When the energy perspective is not taken into account, these devices may fail in critical situations thus leading to the loss of availability of the application and disrupting the QoS of the hosted application.

The work by Ref. [19] has used the advantages of particle swarm algorithm in the optimization of the smart environments consisting of the heating and the cooling systems. This environment as it turns out to be smart tends to utilize more energy, and the author's work has brought about optimization in the office environment. There have been studies in the optimization problems modeling them as a Pareto optimal curve, which states that there are no other feasible solutions that are more optimal than the current solution for the given environment.

The work in [20] has used the advantages of the particle swarm optimization (PSO) algorithm embedded into the support vector machine and use it for energy optimization in the smart cities. Fig. 5.5 gives the overall view of the work done for energy optimization.

In continuation of this research perspective, as the computing power has increased, the use of artificial neural networks (ANNs) has evolved providing better prediction in terms of accuracy. The work by Ref. [21] has used the ANN model for the prediction of daily electricity consumption in Taiwan, which has better accuracy than the traditional model. The research has also revealed that proper attribute selection for the work based on the historical data also influences the efficiency of the algorithm. The use of principal component analysis (PCA) and variants such as the robust PCA have proved to improve the prediction results [22,23].

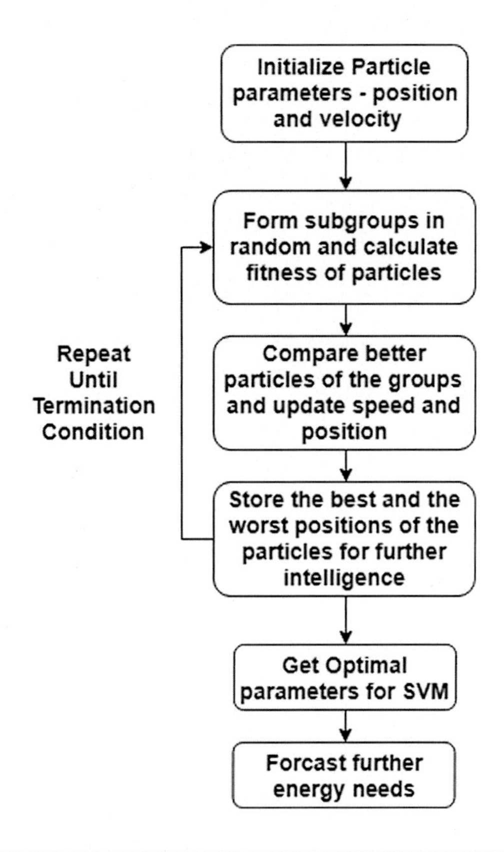

FIGURE 5.5

PSO-based energy prediction.

4.2.2 Artificial bee colony optimization

The research in this perspective started from the era of wireless sensor networks and the need to minimize the energy utilization of the sensors. There have been several studies in the field of energy optimization, and the use of the artificial bee colony (ABC) algorithm has been a prevalent algorithm in the field of optimization. Although there has been already existing heuristics for energy optimization, the ABC algorithm outperforms the traditional algorithms in search ability.

Fig. 5.6 shows the overall working of the ABC algorithm in the prediction of resources to improve the energy utility of the application.

The algorithm involves several steps that starts with the initialization of the parametric values, which include the population number and dimension vector to be optimized, including the lower and upper bounds of each element. Once the parameters have been initialized, the next step involves the initialization of the bee colony, which includes the employer bee, outlooker bee, and the scout bees. Along with this, the fitness estimation is also initialized. The initial step involves the working of the employer bees wherein their solutions are updated based on their fitness values. Now based on the fitness value, the probability values are calculated, which is used by the onlooker bee to find the new solution. This procedure is repeated until

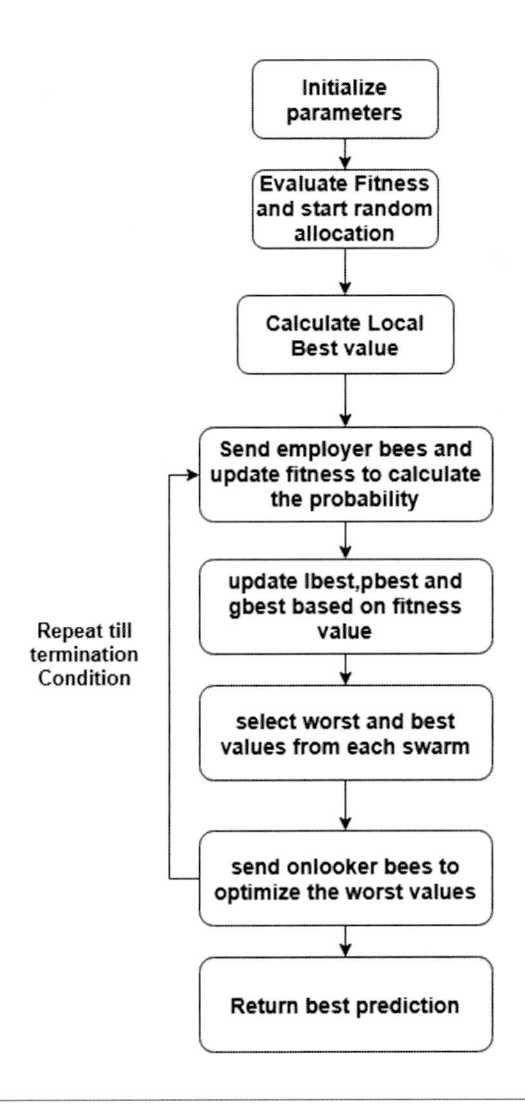

FIGURE 5.6

ABC-based energy prediction.

the maximum cycle number is reached. Finally, the optimal solution is rendered based on the probability values and fitness value.

4.2.3 Bat algorithm-based optimization

As the network scales to support mobile nodes and scalable nodes for any large application, the problem can no longer be modeled as a linear solution and thereby the solution can never be deterministic. Similarly, most of the energy optimization problem can only be solved as an NP-hard problem. The research in the field of NP-hard problems, the evolutionary algorithms also have started evolving, and thus the introduction of the BAT algorithm was introduced to find the optimal short distance between nodes and in selecting optimal cluster head to improve cluster, which can be related to energy-aware load distribution. There have also been parallel versions of the algorithm to improve the performance of the algorithm along with the overall performance of the application.

The advantage of BAT optimization is its rate of convergence. Although the initial solution proposed in the algorithm is based on random values from the bat population, the solution later converges to an optimal point based on the movement of the bats. The basics of the algorithm lie in the chaotic theory, which is an alternate for the random variables of the algorithm. The use of chaotic theory helps in providing higher mobility and variable distribution. One of the powerful chaotic maps is the Sin chaotic map given by Eq. (5.1)

$$x_{k+1} = \frac{a}{4}\sin(\pi x_k) \qquad 5.1$$

where x is the random value to be chaotically chosen and the value lies between 0 and 4 in this case. As the chaotic map result generation depends on the previous results also, the results are relevant and more accurate than choosing a mere random value. With this available literature, our proposed work deals with the proactive method of intelligent energy optimization based on RL.

5. Proposed algorithm—RL-based scheduling

The advantage of the intelligence in the RL is taken up and the proposed algorithm consists of the RL scheduler with the combination of neural networks for approximate scheduling rate. The scheduler presented here is a distributed scheduler, which reduces the algorithmic complexity of the load posed on a single node to run a heavy algorithm. The overview of the algorithm is as follows as given in Fig. 5.7.

The testbed used for the experiments consists of the devices with configuration as given in Table 5.2, which together forms the fog computing environment. Implementation of the infrastructure using a single controller leads to a single point of failure.

The energy consumption of a node is dependent on the CPU and the memory consumption of the node based on the tasks executed. The CPU usage can further be either a static energy utilization or leakage energy when the node is idle or dynamic in nature at times of heavy load execution. Thus, the energy of a node is given as in Eq. (5.2).

$$e_{ij} = k_j * n_{ij} * f_j^2 \qquad 5.2$$

where k_j is the constant of proportionality, n_{ij} is the number of cycles the task takes to execute at the node j, and f_j is the execution frequency, which is the processor speed. Thus, considering these factors, the total energy utilization of the system can be given as in Eq. (5.3)

$$\text{Total energy} = \sum_{i=1}^{m}\sum_{j=1}^{n} e_{ij} \qquad 5.3$$

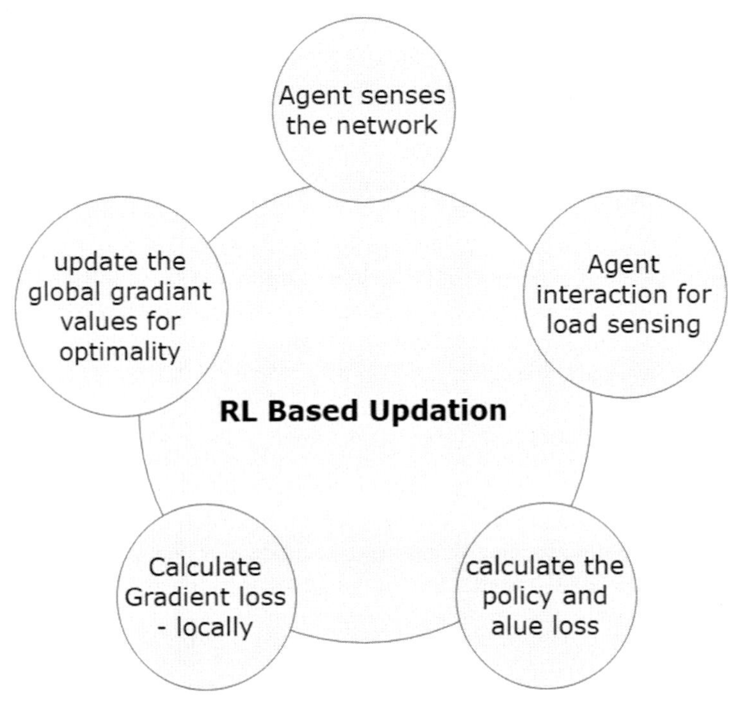

FIGURE 5.7

RL-based intelligent update.

Table 5.2 Design components.

S.No	Description	Purpose
1	Cisco Nexus 5672	Networking
2	Ryu	Fog controller
3	OpenFlow 1.3	Create VRF
4	Multi Router Traffic Grapher (MRTG)	Traffic virtualization
5	Ostinato	Traffic generator
6	IFogSim	Fog simulator
7	Mininet	SDN emulator
8	Docker	Container creation

where m denotes the total number of tasks for execution, and n denotes the total number of nodes in the network. The major resources that decide the performance of the system are CPU, memory, storage, and bandwidth. It is assumed that our architecture and the systems have sufficient memory, storage capacity, and bandwidth for the execution of the incoming tasks. The major concern is the effective utilization of the CPU. Similarly, the processing delay also is an important deciding factor for the performance of the application. The total cost of the system depends on the processing delay, the power consumption, and the migration cost in case of mobility and load balancing. The total cost is given by Eq. (5.4)

$$\text{Total cost} = d_{total} + e_{total} + m_{total} \quad 5.4$$

where d_{total} denotes the total delay that consists of the network and the computation delay, e_{total} is the total energy consumption given in Eq. (5.3), and m_{total} denotes the migration delay in the fog network. The use of containers in the fog nodes facilitates the least migration cost as compared to the VM migration cost. Thus, this cost being minimal can be ignored when compared to the network delay and the energy dissipation.

5.1 Decision on task offloading

The task dispersion to the cloud and the edge devices has to be chosen carefully such that the energy depletion not on a particular node and the latency-sensitive tasks are not pushed to the cloud instead of handling them on the edge. This not only increases the bandwidth utility of the application but also the overall latency of the application.

Thus, the decision of load dispersion to the cloud and the edge has to be systematic and has been done meticulously based on the priority of the tasks received. The decision on offloading is given by Algorithm 1.

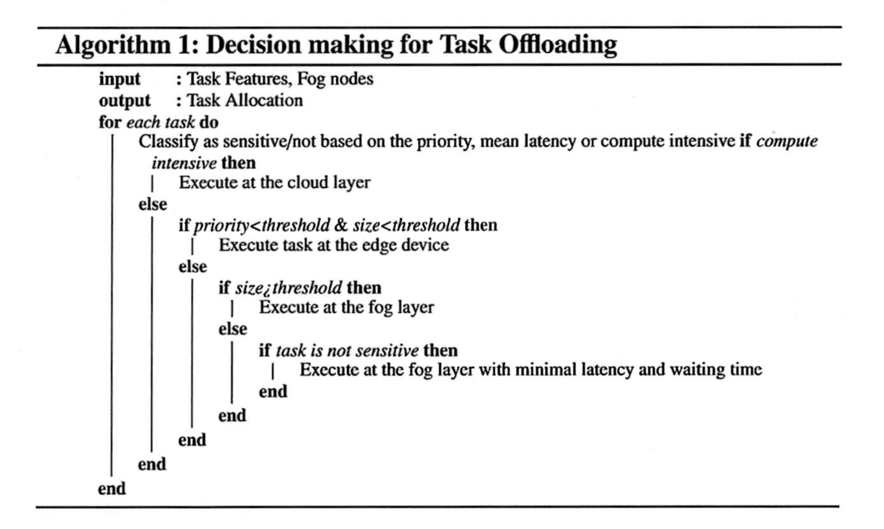

Algorithm 1: Decision making for Task Offloading

```
input   : Task Features, Fog nodes
output  : Task Allocation
for each task do
    Classify as sensitive/not based on the priority, mean latency or compute intensive if compute
      intensive then
        Execute at the cloud layer
    else
        if priority<threshold & size<threshold then
            Execute task at the edge device
        else
            if size¿threshold then
                Execute at the fog layer
            else
                if task is not sensitive then
                    Execute at the fog layer with minimal latency and waiting time
                end
            end
        end
    end
end
```

The offloading of the task is done with consideration of the energy efficiency of the edge nodes and the type of task that is taken up for execution. The algorithm ensures that the time-sensitive applications are given high priority and the timely execution is ensured. The use of RL indorses intelligence in choosing the right actions. The load in the network is captured using the monitoring node, which helps visualize the overall traffic pattern in the network.

With the proposed architecture, the algorithm was put to execution in the created testbed capturing the load and distributing among the nodes based on the energy consideration and the threshold of the load that can be handled by the fog nodes for optimal performance of the application. A huge traffic surge was created to test the proper load balancing among the deployed fog nodes.

5.2 Efficiency of the proposed infrastructure

The fog layer is built on premise, such that the nodes are only at a one-hop distance from the edge devices so that the intermediary computations are done in near real time and reduces the computational latency as compared to sending the data to and from the cloud layer. As the data are sent to a smaller distance and the amount of communication reduces, the power exerted to execute the application is very minimal. As the number of tasks increases, the CPU is utilized as much as possible to increase the efficiency of the fog nodes. The load is intelligently balanced among the nodes in the fog layer such that the energy consumption is taken care and the longevity of the fog nodes is ensured. The simulation of the proposed infrastructure in IFogsim is done such that the efficiency of the fog layer with respect to random load allocation and the intelligent load dispersion is analyzed and compared as given in Fig. 5.8.

The graph showing the dependence of CPU load and power consumption shows that the load balancing helps in the conservation of energy with the help of the optimized payoff and the choice of actions with respect to allocation. Although the power

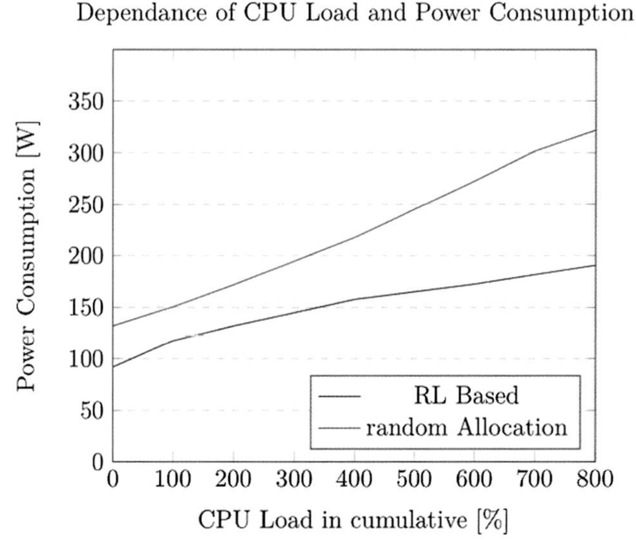

FIGURE 5.8

Dependence of CPU load and power consumption.

consumption increases with load it is much lesser than the random allocation and as the model progresses, the allocation becomes much more intelligent and efficient.

The proper use of decision algorithms for task offloading and the load dissipation based on the energy of the nodes together help in the proper energy conservation and energy-efficient load dissipation of the application.

6. Conclusion and future work

The work in this chapter gives insights on the need for a fog infrastructure as an additive layer to the available cloud computing platform. The chapter also discussed the role of edge/fog computing in the healthcare sector. Fog computing helps in real-time decisions for latency-critical applications in the health sector. The chapter looks into both the architectural and the algorithmic perspective of fog computing. On exploring the available standard architectures for fog, our proposed work also deals with the creation of a real-time testbed with SDN as the networking component to handle the communication among the fog nodes. Moving on to the algorithmic perspective, the literature consists of meta-heuristics for scheduling and there were no stipulated task offloading algorithms defined in the literature. Our proposed work deals with the motive for proactive load dispersion based on the incoming tasks to be executed on the fog layer using RL. In addition, the proposed work deals with a task offloading decision-making of allocating tasks to either the cloud or the fog layer. Finally, the results show that the RL-based algorithm performs better than the random allocation policy in terms of energy conservation.

The future scope of the work lies in the optimization of the RL algorithm and fine-tuning the learning parameters of the algorithm for better performance. The other part of the work also can be extended to fault tolerance of the fog nodes to improve the availability of the resources for the hosted application.

References

[1] Mukherjee M, Matam R, Shu L, Maglaras L, Ferrag MA, Choudhury N, et al. Security and privacy in fog computing: Challenges. IEEE Access 2017;5:19293–304.

[2] Bonomi F, Milito R, Natarajan P, Zhu J. Fog computing: a platform for internet of things and analytics. In: Big data and internet of things: a roadmap for smart environments. Springer; 2014. p. 169–86.

[3] DoukasC MI. Bringing iot and cloud computing towards pervasive healthcare. In: 2012 Sixth international conference on innovative mobile and internet services in ubiquitous computing. IEEE; 2012. p. 922–6.

[4] Biswas AR, Giaffreda R. IoT and cloud convergence: opportunities and challenges. In: 2014 IEEE world forum on internet of things (WF-IoT). IEEE; 2014. p. 375–6.

[5] Mahmud R, Ramamohanarao K, Buyya R. Latency-aware application module management for fog computing environments. ACM Transactions on Internet Technology (TOIT) 2018;19(1):9.

[6] HongK LD, Ramachandran U, Ottenwälder B, Koldehofe B. Mobile fog: a programming model for large-scale applications on the internet of things. In: Proceedings of the second ACM SIGCOMM workshop on Mobile cloud computing. ACM; 2013. p. 15–20.
[7] Yangui S, Ravindran P, Bibani O, Glitho RH, Hadj-Alouane NB, Morrow MJ, et al. A platform as-a-service for hybrid cloud/fog environments. In: 2016 IEEE international symposium on local and metropolitan area networks (LANMAN). IEEE; 2016. p. 1–7.
[8] Bitten court LF, Lopes MM, Petri I, Rana OF. Towards virtual machine migration in fog computing. In: 2015 10th international conference on P2P, parallel, grid, cloud and internet computing (3PGCIC). IEEE; 2015. p. 1–8.
[9] Agarwal S, Yadav S, Yadav AK. An efficient architecture and algorithm for resource provisioning in fog computing. International Journal of Information Engineering and Electronic Business 2016;8(1):48.
[10] Al Faruque MA, Vatanparvar K. Energy management-as-a-service over fog computing platform. IEEE Internet of Things Journal 2015;3(2):161–9.
[11] Ravi N, Selvaraj MS, Tefens. Testbed for experimenting next-generation-network security. In: 2018 IEEE 5G world forum (5GWF). IEEE; 2018. p. 204–9.
[12] TranMQ, NguyenDT LVA, NguyenDH PTV. Task placement on fog computing made efficient for IoT application provision. Wireless Communications and Mobile Computing 2019;2019.
[13] Ray PP, Dash D, De D. Edge computing for internet of things: a survey, e-healthcare case study and future direction. Journal of Network and Computer Applications 2019;140:1–22.
[14] Sels V, Gheysen N, Vanhoucke M. A comparison of priority rules for the job shop scheduling problem under different flow time-and tardiness-related objective functions. International Journal of Production Research 2012;50(15):4255–70.
[15] Wang S, Urgaonkar R, Zafer M, He T, Chan K, Leung KK. Dynamic service migration in mobile edge-clouds. In: 2015 IFIP networking conference (IFIP networking). IEEE; 2015. p. 1–9.
[16] Razzaq S, Wahid A, Khan F, ul Amin N, Shah MA, Akhunzada A, Ali I. Scheduling algorithms for high-performance computing: an application perspective of fog computing. In: Recent trends and advances in wireless and IoT-enabled networks. Cham: Springer; 2019. p. 107–17.
[17] Gong X, Deng Q, Gong G, Liu W, Ren Q. A memetic algorithm for multi-objective flexible job-shop problem with worker flexibility. International Journal of Production Research 2018;56(7):2506–22.
[18] Milan ST, Rajabion L, Ranjbar H, Navimipoir NJ. Nature inspired meta-heuristic algorithms for solving the load-balancing problem in cloud environments. Computers and Operations Research 2019;110:59–187.
[19] Kusiak A, Xu G, Tang F. Optimization of an HVAC system with a strength multi-objective particle-swarm algorithm. Energy 2011;36(10):5935–43.
[20] Zhang L, Ge R, Chai J. Prediction of China's energy consumption based on robust principal component analysis and PSO-LSSVM optimized by the Tabu search algorithm. Energies 2019;12(1):196.
[21] Meng F, Liu Y, Liu L, Li Y, Wang F. Studies on mathematical models of wet adhesion and lifetime prediction of organic coating/steel by grey system theory. Materials 2017; 10(7):715.

[22] Van Luong H, Deligiannis N, Seiler J, Forchhammer S, Kaup A. Compressive online robust principal component analysis via backslash minimization. IEEE Transactions on Image Processing 2018;27(9):4314–29.
[23] Chrétien S, Clarkson P, Garcia MS. Application of robust PCA with a structured outlier matrix to topology estimation in power grids. International Journal of Electrical Power and Energy Systems 2018;100:559–64.

CHAPTER

Deep learning-based security schemes for implantable medical devices

6

Heena Rathore, PhD, BE [1], Amr Mohamed, PhD [2], Mohsen Guizani[2]

[1]*Department of Computer Science, University of Texas, San Antonio, TX, United States;* [2]*Professor, Department of Computer Science and Engineering, Qatar University, Doha, Qatar*

1. Introduction

The need for higher effectiveness and patient mobility has accelerated the fast development of wireless technologies and applications in the healthcare sector, especially in hospitals and medical centers. Medical devices, as an example, are increasingly dependent on connectivity to the Internet for on-demand therapy control and transmission of medical data to and from a cloud-based server. This aids in real-time monitoring without the patient being physically present with the doctor or the caregiver. This is highly beneficial for in-patient comfort while preserving and securing their information.

Implantable medical devices are typically tiny in size and are installed either inside or outside human patients' body for treating a diverse set of diseases and ailments. Examples of some of these devices are shown in Fig. 6.1. The fast growth of medical devices equipped with the ability to store and communicate medical data and adjust treatment plans on-demand remotely has raised critical security and privacy concerns. There are several affirmed reports of standard computer infections contaminating medical devices in radiology, heart catheterization research facilities, sleep labs, and other clinical divisions [2]. Furthermore, Jay Radcliffe has shown how an insulin pump can be manipulated to disperse a lethal amount of insulin [3]. There is a focused effort across research groups to develop a standard to harmonize technical specifications, rules, methods, and definitions related to the security of medical devices, and deep learning is one of the techniques being actively considered.

In this era of big data fueled by the growth of social media content and billions of connected devices, an enormous amount of information is being generated and managed. This has led to the (re)emergence of artificial intelligence (AI), with machine learning being one of the technologies that are fueling this comeback. With this massive growth in the amount of data that can be analyzed, the presence of a human in the middle for the feature extraction phase is turning out to be the weak

Energy Efficiency of Medical Devices and Healthcare Applications. https://doi.org/10.1016/B978-0-12-819045-6.00006-6

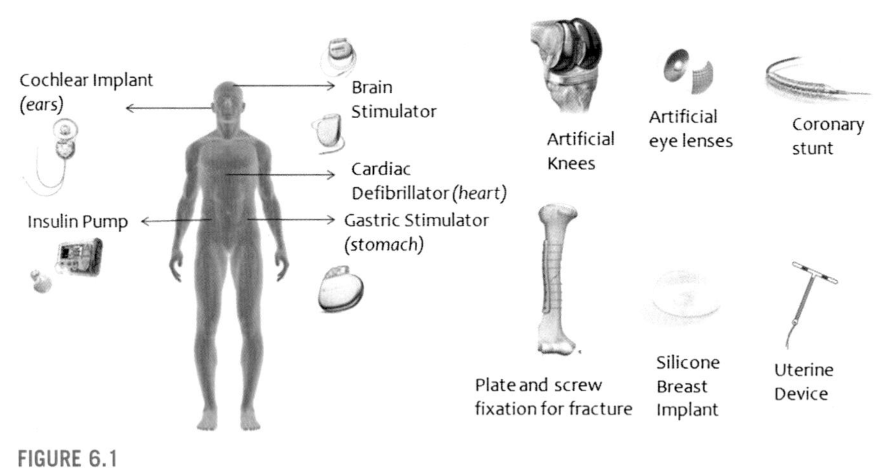

FIGURE 6.1

Eleven most widely used wireless/implantable medical devices.

link in the chain. The advent of faster graphical, embedded, and tensor processors is shining the spotlight on this weakness even more prominently. This job is now being delegated to computers via an emerging technology known as deep learning, which attempts to imitate how the human brain works, albeit at a much larger scale. Computers have shown to be able to handle large amounts of data well through innovations in computing and storage technologies. This capability is quickly finding its way in applications such as connected cars and smart factories.

Deep learning, based on artificial neural networks, continuously adapts the machine's behavior from learning algorithms and leveraging the constantly growing amount of data increases the efficiency of training processes. Deep learning provides incredible benefits to a diverse set of problems such as classification, segmentation, and feature engineering.

Our research effort is driven by the motivation to use deep-learning algorithms for securing cyberphysical systems such as implantable medical devices (IMDs). By doing so, we hope to address the privacy and security breaches in IMDs, which have alarmed both the health providers and government agencies. Our work is instrumental in ensuring security to these small devices, which is a vital task to prevent severe health consequences to the bearer. To achieve this goal, we have designed novel algorithms, such that they are practical to implement on real-world devices. As described earlier, IMDs use sensing, communication, and control capabilities to treat the patient's health and give a mechanism to provide regular remote monitoring to the healthcare providers. However, current wireless communication channels can curb the security and privacy of such devices by allowing an attacker to interfere with both the data and communication channels. Such attacks can range from system to infrastructure levels where both the software as well as the hardware of the IMD can be compromised. In recent years, biometric and cryptographic approaches for authentication, machine-learning approaches for anomaly detection, and external wearable devices for wireless communication protection have been proposed. However, the existing solutions for wireless medical devices are either computationally expensive for memory-constrained devices or require additional devices to be worn. To overcome this challenge, we have proposed effective and secure data communication by introducing policies that will incentivize the development of security techniques. Deep learning is among the most effective and broadly utilized techniques for classification, identification, and segmentation. Although they are effective, they are both computationally and memory intensive, making them hard to be deployed on low-power embedded frameworks. Our proposed work presents an on-chip neural system for securing diabetic, cardiovascular, and Parkinson's disease treatments. Experimental results prove that our work is more efficient than optimized systems, as it is capable of self-learning, self-healing, and are self-adaptive in nature. Our proposed model is not only less cumbersome in comparison to optimized solutions but also is quite efficient and robust. It also proves to be resilient toward critical errors. Overall, we hope that each of our contributions leads to an improved understanding of networks and efficient modeling that permit system operations to function at a higher level.

This chapter focusses on how deep-learning and other machine-learning algorithms can be utilized for making medical devices more secure and efficient. Section 2 provides the mapping of deep learning with the human brain. Section 3 provides details on the technical concepts behind deep learning. Section 4 provides a review of security challenges, attacks, and resolutions for wireless medical devices. The chapter is divided into providing security at the sensor data level, communication level, and application level. These are discussed in Sections 5, 6, and 7, respectively. Section 8 concludes the chapter.

2. Mapping of deep learning and human brain

Deep learning is not just a technique that can revolutionize artificial intelligence but is something fundamentally like the human brain. Although as often reported, it is true that there are similarities between the human brain and deep learning, one should quickly acknowledge that there are areas in which one outperforms the other. To understand this better, let us first review how the human brain works. The human brain contains billions of nerve cells organized in the form of a network, which coordinates thoughts, emotions, sensation, and movement, to produce a behavior. One can think of this as a mechanism that allows transmission of electrochemical signals in a fraction of a second, through a complex highway system of nerves connecting the brain with the rest of the body.

The most common visual representation of the brain is the deep fold and wrinkles on the cerebral cortex (outer layer of the cerebrum) where the information is processed [1]. The corpus callosum, a thick tract of nerves, is located at the base of the deep fissure that separates the two hemispheres to constitute the cerebrum, which is the primary channel for communication. These hemispheres are divided into four lobes that work in tandem with structures, which come in pairs and are known as the limbic system that controls emotions and memories. The functions related to critical thinking, thought arrangements, and transient memory management are handled by the frontal lobes. Sensory information related to taste, temperature, and touch is handled by the parietal lobes. The job of processing images captured by our eyes and comparing them to prior experience is managed by the occipital lobes. The temporal lobes manage memory retention by processing data from our sensory organs responsible for sound, taste, and smell. The cerebellum is located underneath and behind the remainder of the cerebrum and is responsible for facilitating development through the consolidation of tangible data from the eyes, ears, and muscles. The hippocampus sends recollections to be put away in fitting areas of the cerebrum and after that reviews them when essential. The peripheral nervous system provides a communication pathway between the mind and limits and contains each of the nerves in the body, aside from the ones in the cerebrum and spinal cord. For example, your cerebrum tells the muscles in your arm and hand to remove your finger off the hot stove in a fraction of a second. Dendrites and axons are the two types of branches falling off the nerve cells. The former receives incoming messages from other nerve cells, whereas the latter perform the contrary function. A single nerve cell (neuron) is connected to other neurons in the body and they communicate with each other through electrical impulses when the nerve cell is stimulated. Inside a neuron, the impulse moves to the tip of an axon and causes the release of synapses and synthetic compounds that act as delegates neurotransmitters. Synapses go through the neural connection (which is the gap between two nerve cells), thereby appending the receptors to the receiving cell. As these synapses travel to their destination, they allow us to manage movement, thoughts, and emotions.

An IMD, as a specific example of cyberphysical system, consists of actuators, sensors, a controller, and a plant, as shown in Fig. 6.2.

If the brain were to be modeled as a cyberphysical system, the four different lobes that make up the hemisphere, namely the frontal, parietal, occipital, and

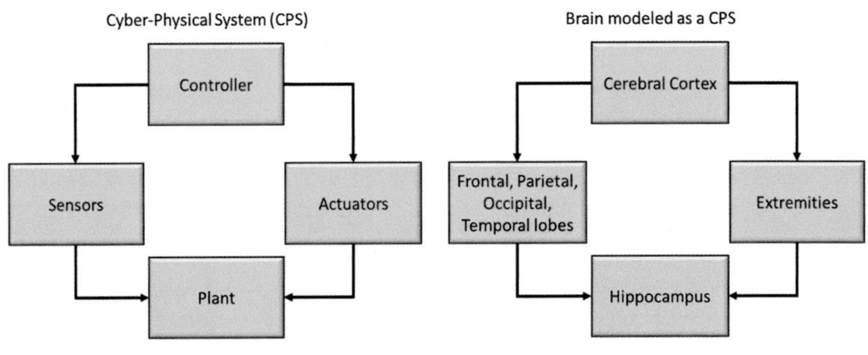

FIGURE 6.2

Mapping of IMDs or CPSs to brain functions.

temporal lobes, act as the sensory elements of this CPS. They receive signals from various sensing organs, such as nose, ears, and eyes as well as prestored memories in the hippocampus. The cerebral cortex does the job of information processing. The cerebellum also plays a pivotal role in this aspect. Decisions made in the cerebral cortex and cerebellum then coordinate the movement in different extremities, such as hands and fingers. Dendrites and axons perform these functions. The hippocampus sends memories to be stored in the brain. When a child is born, the hippocampus is yet to be developed. The child learns through different experiences and acquired wisdom from others around her, on the type of responses for different stimuli. As the child grows, the hippocampus matures storing judgments and prior emotions. These play a pivotal role in deciding the type of response for a stimulus. In machine learning, the role of the hippocampus is accelerated via the use of training data. Any system uses prior data to train rapidly. There are examples of applications such as self-driving cars, where the car is in an experimentation phase to learn as it is driving around.

In many ways, an IMD closely resembles a brain in its ability to sense external signals, make decisions based on the sensing, and then make command decisions. This has led researchers to look at the brain as an inspiration to manage various aspects of cyberphysical systems, including the security aspects. Deep learning is modeled after an artificial neural network similar to a system present in the human cerebrum. The "deep" in "deep learning" originates from the number of layers in the network of artificial neurons. As information goes through this artificial network, each layer forms a part of the information that can detect anomalies and spot predictable patterns, thereby yielding the desired output.

In a nutshell, deep learning relies on many layers of virtual neurons in the network to achieve sophisticated kinds of learnings. The network is structured into hierarchies, each trying to achieve a specific objective. Let us take number detection as an example. The vertical and horizontal strokes that make up a number are processed by the lower-level neurons, whereas more abstract aspects such as position in the sequence and related operations are handled by the higher-level neurons. The network uses repeated trial-and-error techniques to learn from the initially supplied datasets, known as training data. Through this iterative process, the algorithm computes the error at the output at each layer and provides it as feedback so that each layer can take steps to minimize the error.

3. Neural network architecture

The brain is considered as the computer of human beings. In deep learning, with the help of neural networks, computer programs are assembled from millions of artificial brain cells (neurons) that learn and evolve in a strikingly comparable manner to human minds. With the set of input, the system learns and updates the internal state that in turn helps in predicting the inputs using the learning accomplished by internal state. The fundamental unit in neural networks is the neuron and is structured to form three layers namely input, hidden, and output layer as shown in Fig. 6.3.

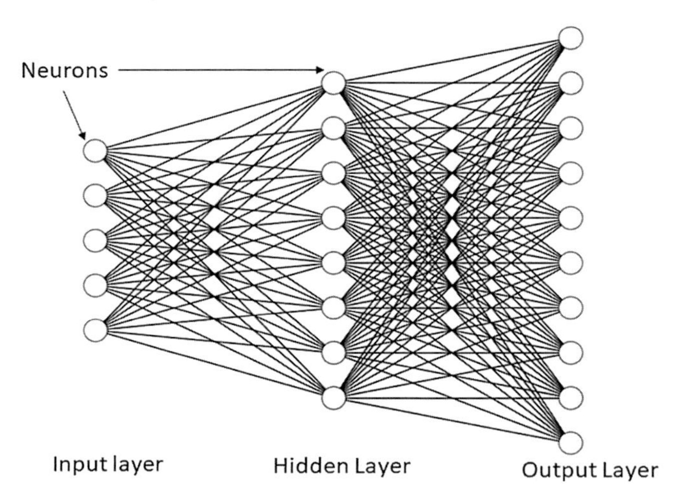

FIGURE 6.3

Neural network architecture.

Neural networks (deep learning) in general has basic seven steps:

1. *Model initialization*: Each neuron receives a set of inputs biased with a certain weight. The neuron then uses its activation function to calculate the output. Consider a situation, wherein the activation function is a linear combination of inputs, as shown in Eq. (6.1)

$$Y = w.x \tag{6.1}$$

where Y, x, and w represent the output, the input, and the weight, respectively, as shown in Fig. 6.4.

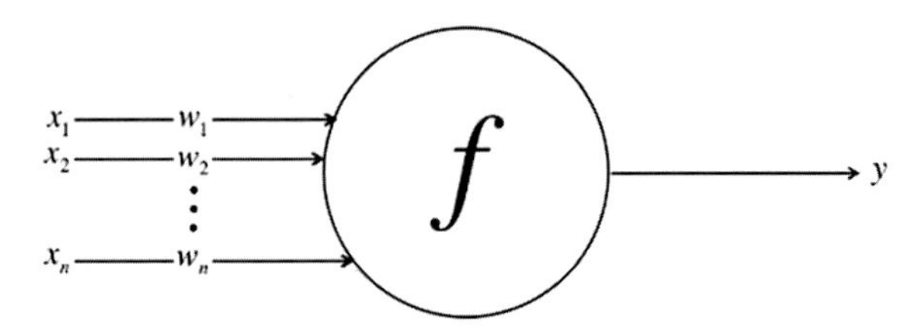

FIGURE 6.4

Model initialization schematic.

2. *Forward propagation*: As the name implies, the input information is fed in a forward way through the system to generate the output. Each hidden layer acknowledges the input information, processes it according to the activation function, and forwards it to the next layer. At each hidden or output layer, the processing happens in two steps: preactivation and activation (shown in Fig. 6.5). In the preactivation phase, the weighted sum of inputs is computed.

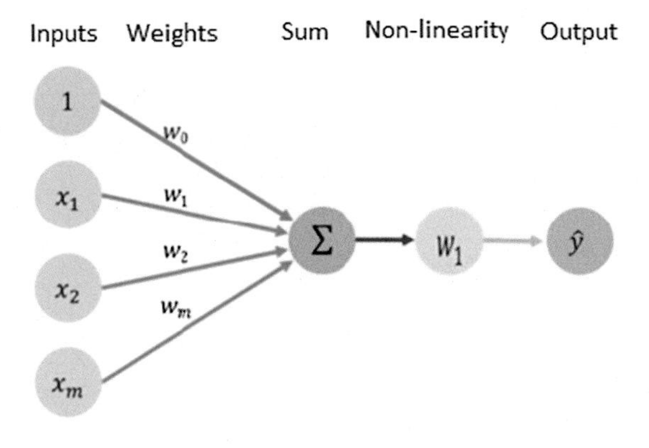

FIGURE 6.5

Node of a neural network.

This is done to decide whether the information can be passed forward or not. Followed to preactivation function, an activation function, examples of which include sigmoid, tanh, ReLU, and Softmax, is added. The main reason why we use the sigmoid function is that it exists between 0 and 1. It is especially used for models where we have to predict the probability as an output. As the probability of anything exists only between the range of 0 and 1, sigmoid is the right choice. The range of the tanh function is from −1 to 1. tanh is also sigmoidal (s-shaped). The advantage is that the negative inputs will be mapped strongly negative and the zero inputs will be mapped near zero in the tanh graph. The tanh function is mainly used for classification between two classes. RELU is a piecewise linear activation function. The purpose of the activation function is to introduce nonlinearities into the network. In addition, bias is added to shift the activation function either left or right.

3. *Loss function*: To calculate the model error between the predicted value (y') and the actual value (y), the loss function is computed. There are different ways to calculate the loss errors, and these are mean squared error, mean squared logarithmic error, l2, mean absolute error, mean absolute percentage error, etc. An example of mean square error calculation is

$$loss = \frac{1}{n} \sum_{n}^{i=1} \left(y^{(i)} - y'^{(i)} \right)^2 \tag{6.2}$$

4. *Differentiation*: Differentiation is used to modify the internal weights of neural networks with the goal of minimizing the previously defined total loss function. The derivative of a function at a certain timestamp is a mathematical representation of the rate at which the function is changing as a function of time. If the derivative is positive, it implies that the function is increasing and vice versa if the derivative is negative. We adjust the value of the weights to counteract this

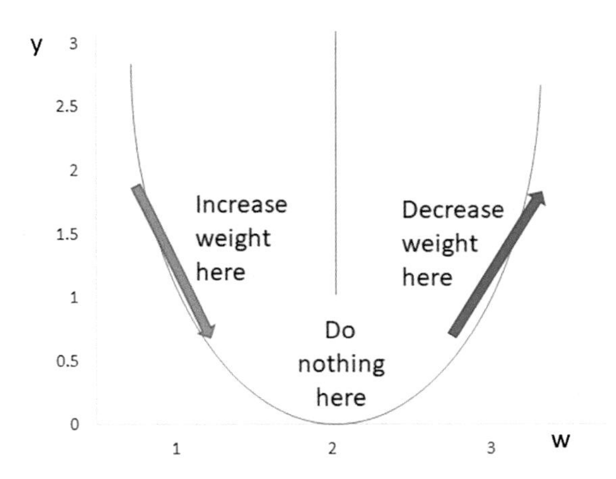

FIGURE 6.6

Differentiation function toward a minimum through weight adjustments.

behavior. After a certain number of iterations, the derivative stabilizes to zero, thereby indicating stability.

5. *Backpropagation*: A neural network propagates the input information forward through its parameters toward the moment of decision and after that back-propagates information about the error, in backward direction through the system, with the goal that it can change the parameters.
6. *Weight update*: The weights are updated using Eq. (6.3):

$$nw = ow - d * r \tag{6.3}$$

where *nw* is the new weight, *ow* is the old weight, *r* is the learning rate, and *d* is the derivative rate that is calculated from the differentiation step

7. Iterate until convergence: This process can take several iterations to converge. In each of these iterations, the weights are either increased or decreased in small steps, such that they converge to a minimum value of the differentiation function, as shown in Fig. 6.6.

4. Security of implantable medical devices

Currently, the largest market in the world for medical devices is the United States. As per the report [4], worldwide dynamic implantable medical devices market was esteemed at roughly USD 16.47 billion in 2017 and is forecasted to create an income of around USD 23.33 billion before the end of 2024, developing at a compound annual growth rate of around 5.10%. Needless to say, securing such massive medical devices is a necessity.

Technology advancement in health management has not only added administrative services to well-being and medical centers but also improved usability and accessibility. Progressive advancement of IMDs has affected human lives, not only by bringing improvement in the healthy lives, but has also included new risks. In the late 90s, six deaths were reported in close succession due to assembling or programming errors in the Therac-25 device [11]. With the advent of IMDs in the early 20s, few failures and issues were raised by the usage of these medical gadgets. The tragic death of a youth due to a short circuit in her pacemaker is one of the early documented incidents of the catastrophic impacts of such failures [12]. Additionally, users of insulin pumps have reported attacks in which intruders illegally listened to wireless communications with the intent of maliciously controlling the gadgets. Some examples of this include turning insulin pump on or off and infusing wrong dosages of insulins nonprescribed number of times. Nevertheless, there were reports where the patients died after receiving diathermy therapy for implanted deep brain stimulators. Similarly, gastric simulators were made to fail by presenting them with imperfect electrical signals thereby simulating false gastric emptiness situations. IMDs should be made secure against such unintended consequences of the technology advances, which are designed to improve the quality of medicine for patients.

The first of such security mechanisms aims at avoiding intrusion through well-known methods such as the one using biometric fingerprints, the distance between the patient and the device, strategic management of keys and audit mechanisms. The second one aims at detecting intrusion by spotting differences between prescribed and actual behaviors [7–9,20]. Biometric approaches consider omnipresent attributes related to eyes [13] or heart [15] to provide verification and validation. Although such attributes are always present, they tend to evolve over time, a trait that is not managed well by such approaches. Additionally, such methods are easily influenced by user training, sample gathering, and other natural environmental conditions. Distance-based approaches use radar-like techniques to measure the time delay between transmission and reception to estimate the distance between the IMD and caregiver [14]. It then uses this information to authenticate a valid device. The near-field communication techniques proposed in Refs. [16,27] provide weak authentication as the hacker can come closer to the vicinity of the patient. Inspired by public–private key technique, the key management protocol exchanges keys to provide verification and validation between patient and caregiver. However, such protocols [14,17,18,23] could suffer from attacks before the exchange of keys (known as a 0-day attack) and are computationally intensive. Data reports of completed audits [21] are sometimes stored in the IMD for monitoring the information exchanges that have happened in the gadget; however, such techniques are memory intensive. IMDGaurd [28], MedMon [19], IMDShield [25], and Cloaker [24,26,26] are examples of external device methodologies. However, as stated in the name of this technique, they require extra gadgets to be carried beside the wireless medical device. Additionally, they also consume the battery life of the medical device and one can easily breach security by moving in close proximity. Anomaly detection mechanism, as explained in Ref. [22], classifies between the regular and irregular patterns and learns the network in the way to alarm whenever a suspicious activity is observed. Anomaly-based approaches exhaust resources.

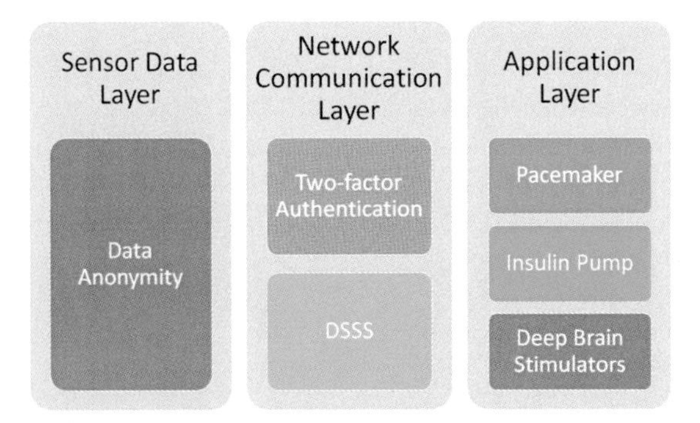

FIGURE 6.7

Security measures at different layers of IMD.

Security of IMDs needs to be handled at each of the three layers, namely sensor layer, communication layer, and application layer, which form the technical framework of the platform for such devices. Each layer has its own core functions, and each layer works with the other to form the basic architecture of the medical platform as shown in Fig. 6.7. The chapter covers security techniques for the following:

- Sensory data layer: Security in ensuring the correctness of the data being acquired by the sensors.
- Communication layer: Securing the communication link through authentication and encryption to avoid tampering.
- Application layer: Techniques for in-device detecting intrusions for a diverse set of medical devices such as insulin pumps, deep brain stimulators, and pacemakers.

To highlight the categories of security techniques, in the rest of the chapter, we will talk about (1) anonymizing data (through anonymity techniques), (2) security of the communication link (through direct sequence spread spectrum (DSSS), and (3) in-device intrusion detection (e.g., machine learning for diabetes).

5. Sensor data layer security

Anonymization is the technique of removing an individual's direct and indirect identifiers that may prompt an individual being distinguished. An individual's information such as their name, address, zip code, or phone number can be used to uniquely distinguish them from another person. Similarly, certain data related to the work environment, such as job title or pay, can be used to indirectly distinguish an individual. In either case, anonymization helps in protecting the personal data that, in turn, assists in avoiding the identification of an individual. The definition of personal data in general data protection regulations includes information such as login details, age, and address.

There are different ways of anonymizing the data, which include randomization, and generalization. Randomization employs different techniques such as noise addition, substitution/permutation, and differential privacy.

- Noise addition: Here the personal identifiers are expressed by adding some noise for instance weight: 120 lbs to >145 lbs.
- Substitution/permutation: Here, either the position or the values of the personal identifiers are randomized within a table, for instance: Phone: 10120 is replaced by Phone: telephone
- Differential privacy: Here individual identifiers of one information set are compared against an anonymized dataset held by a third party with directions to utilize the noise function and an amount of data leakage is defined and characterized.

Generalization has two well-known techniques, that is, k-anonymity and L-diversity, where personal identifiers are anonymized by distributing them into a range or group, for instance, weight: 120 lbs to >110–125 lbs. In L-diversity, personal identifiers are first summed up, each point within equality class is conditioned such that it occurs at least *n* times, for example, a property is assigned to individual identifiers and that property is made to occur a base number of times with a dataset or a segment.

ARX as a tool for anonymization was used on the diabetic patient dataset partially (i.e., some of the quasi-identifiers fields but not all). ARX anonymization tool supports various anonymization techniques, methods for analyzing data quality, and reidentification risks, and it supports well-known privacy models, such as k-anonymity, l-diversity, and t-closeness.

Anonymization was evaluated on the diabetes dataset obtained from the UCI machine learning repository [29]. A snapshot of the dataset is given in Fig. 6.8.

The task was to perform classification involving the prediction of the genuine or fake glucose dosage in insulin pump therapy. These 10 features included in the dataset were as follows:

- Date: Measurement date
- The patient's age in years
- Plasma glucose concentration measured every 2 hours via an oral glucose tolerance test
- Time: Measurement time
- Diastolic blood pressure measurements in the units of mm Hg
- A serum insulin measurement performed every 2 hours in the units of mu U/mL
- Body mass index (weight in kg/(height in m)^2)
- A function modeling the diabetes pedigree of the patient
- Value: Glucose measurement

Results of the experiments run on datasets (quantity of 12,000) are discussed next. Moreover, 60% of the dataset was used during the training (learning) phase and the rest 40% was used during the testing phase. The number of hidden layers

	A	B	C	D	E	F	G	H	I	J	K
1	Age	Plasma glu	Diastolic I	Triceps sk	2-Hour ser	Body mas:	Diabetes	Date	Time	Value	Label
2	46	118	72	19	0	23.1	1.476	32337	1320	196	1
3	46	100	78	25	184	36.6	0.412	32337	1080	165	0
4	46	134	80	37	370	46.2	0.238	32337	720	144	0
5	46	102	76	37	0	32.9	0.665	32337	480	81	0
6	46	105	90	0	0	29.6	0.197	32338	1080	242	0
7	46	61	82	28	0	34.4	0.243	32338	480	133	0
8	46	84	72	31	0	29.7	0.297	32338	720	190	0
9	46	115	72	0	0	28.9	0.376	32338	480	26	0
10	46	92	62	32	126	32	0.085	32338	1320	251	1
11	46	100	84	33	105	30	0.488	32339	720	180	0
12	46	144	82	46	180	46.1	0.335	32339	480	26	0
13	46	67	76	0	0	45.3	0.194	32339	480	140	0
14	46	155	62	26	495	34	0.543	32339	1320	201	1
15	46	102	76	37	0	32.9	0.665	32339	1080	259	0
16	46	118	72	19	0	23.1	1.476	32340	720	175	0
17	46	100	78	25	184	36.6	0.412	32340	480	26	0
18	46	134	80	37	370	46.2	0.238	32340	480	108	0
19	46	115	72	0	0	28.9	0.376	32340	1320	97	1
20	46	61	82	28	0	34.4	0.243	32340	1080	64	0

FIGURE 6.8

Original diabetes dataset.

were 2. All the features were anonymized as shown in Fig. 6.9, and the results seem quite analogous to the ones achieved without anonymizing the dataset. It achieved 96.63% accuracy with 1.82% standard deviation as opposed to 96.85% accuracy with the unanonymized dataset (Fig. 6.8 Original Dataset).

	A	B	C	D	E	F	G	H	I	J	K
1	Age	Plasma glucose concentration a 2 hours in an oral glucose tolerance test	Diastolic blood pressure	Triceps skin fold thickness	2-Hour serum insulin	Body mass index	Diabetes pedigree function	Date	Time	Value	Label
2	[21, 25[	84	0	0	0	[0, 10[	0.304	33349	549	[0, 50[	0
3	[21, 25[	102	75	23	0	[0, 10[	0.572	33350	820	[0, 50[	0
4	[21, 25[	84	0	0	0	[0, 10[	0.304	33358	1320	[0, 50[	0
5	[21, 25[	102	75	23	0	[0, 10[	0.572	33359	734	[0, 50[	0
6	[21, 25[	84	0	0	0	[0, 10[	0.304	33367	1080	[0, 50[	0
7	[21, 25[	102	75	23	0	[0, 10[	0.572	33368	1050	[0, 50[	0
8	[21, 25[	84	0	0	0	[0, 10[	0.304	33377	582	[0, 50[	0
9	[21, 25[	102	75	23	0	[0, 10[	0.572	33378	720	[0, 50[	0
10	[21, 25[	102	75	23	0	[0, 10[	0.572	33387	1015	[0, 50[	0
11	[21, 25[	84	0	0	0	[0, 10[	0.304	33395	733	[0, 50[	0
12	[21, 25[	118	64	23	89	[0, 10[	1.731	33398	1020	[0, 50[	0
13	[21, 25[	102	75	23	0	[0, 10[	0.572	33403	1031	[0, 50[	0
14	[21, 25[	84	0	0	0	[0, 10[	0.304	33411	737	[0, 50[	0
15	[21, 25[	84	0	0	0	[0, 10[	0.304	33419	1041	[0, 50[	0
16	[21, 25[	102	75	23	0	[0, 10[	0.572	33420	753	[0, 50[	0
17	[21, 25[	84	0	0	0	[0, 10[	0.304	33427	427	[0, 50[	0
18	[21, 25[	102	75	23	0	[0, 10[	0.572	33428	428	[0, 50[	0
19	[21, 25[	118	64	23	89	[0, 10[	1.731	33430	758	[0, 50[	0
20	[21, 25[	84	0	0	0	[0, 10[	0.304	33435	1026	[0, 50[	0

FIGURE 6.9

Anonymized diabetes dataset.

The work presented in this section is significant in various aspects. Sensors in a cyberphysical system can be compromised in various ways, such as interfering with the calibration coefficients stored on the sensor or physically hacking the device itself. The techniques presented in this section use feature set such as measurement date, patient's age, BMI, and blood pressure, to identify if the device has been tampered with. As can be seen, some of these data, specifically the patients' age can be subject to privacy concerns. The second significance of this work is the anonymization of the data, which addresses the privacy concerns, but does not significantly alter the accuracy as compared to the nonanonymized data.

6. Communication layer security

The communication layer in an implantable medical device can be modeled using the TCP-IP model as shown in Fig. 6.10. In this section, we will discuss the different types of attack modes possible at each of the layers. We will describe the DSSS technique as one potential defense strategy to protect the communication layer against attacks.

An IMD selects transmission parameters, such as carrier frequency, data rate (translated to bandwidth), modulation type, and reference level, at the physical layer. IMDs can suffer from a jamming attack at the physical layer, in which an intruder tries to transmit illegally in the same frequency band as the IMD, generally at an elevated power level, thereby messing up with the primary communication. Such type of jamming can either be continuous or distributed on a temporal scale. The latter can prove even more disastrous because it is intermittent in nature and hard to protect against. IMDs may also suffer from a tampering attack in which the

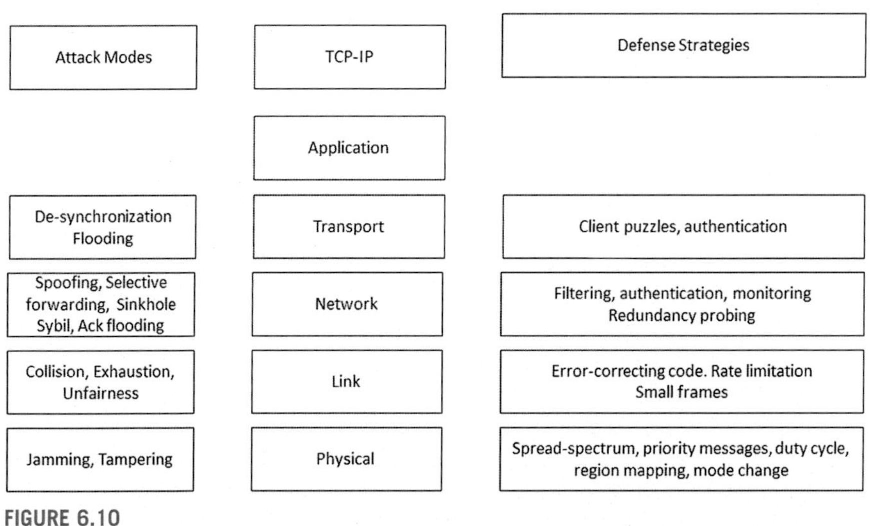

FIGURE 6.10

Attack modes and defense strategies at the communication link.

contents of the signal may be modified or eavesdropped on to be played out repeatedly. In the IMD, the link layer typically manages functions such as multiplexing of data streams, detecting data frames and controlling access to the medium. An intruder can introduce attacks such as intentional collisions, resource exhaustion, and unfair allocation at this layer. Similar to the jamming threat at the physical layer, when two nodes attempt to transmit at the same frequency, their messages can collide. This causes the network to retransmit the packets, thereby causing delays and inefficient bandwidth usage. Typically, such attacks are targeted at special messages such as acknowledgment messages. Resource exhaustion happens when the intruder tries to flood the network with repeated messages to generate collisions. Unfair allocation causes some nodes to be given higher priority over others. An intruder can directly modify and replay or pretend to be the original owner of the routing information at the network layer resulting in traffic disruption in the network. Additionally, he can create recursive routing loops, disrupt traffic from selected nodes, and artificially alter the length of source routes. The network layer is responsible for forwarding messages from all nodes in a multihop network. An intruder can cause selective forwarding, wherein messages from some nodes are not transmitted. Finally, a Sybil attack is the one in which a node may present dual identity in the network and is an effective technique to counter attacks on distributed data storage, resource allocation, and data aggregation. Sinkhole, Wormhole, and Blackhole are other types of network layer attacks in which a network route path is artificially altered by either dropping, introducing, or blocking invalid nodes in the path. The aim of the malicious node could be to change and obstruct the path-finding process or to catch all information being sent to the destination node concerned. Although the earlier-mentioned attacks are typically managed by a single intruder, you can also run into situations where a group of nodes can collude together to flood the network. Such attacks are known as Byzantine attack. They are very hard to recognize, as under such attacks the system does not show anomalous conduct. In any of these attacks, compromised nodes can share insensitive information to unauthorized nodes or deplete network resources such as battery power, bandwidth, and computational power. They may also lead to unauthorized acknowledgments, routing table overflow, routing table poisoning, packet replication, route cache poisoning, and rushing attacks. Finally, an intruder can introduce attacks at the transport layer such as flooding and desynchronization attacks. In the former, memory exhaustion makes it difficult for a protocol to maintain state as the intruder repeatedly makes new connection requests. Finally, an adversary may, for example, over and again send farce messages to an end host making the host demand the retransmission of missed frames, resulting in desynchronization. This can cause wasted energy to recoup from the errors that never truly exist.

The IMD can be defended against each of these attacks in different ways, as illustrated in Fig. 6.11. In this chapter, we specifically refer to the work done in Ref. [6], where an ECG signal is used as a biometric marker for identification. However, as the QRS component in the ECG does not change as a function of time, it is easy to be tampered with at the physical layer. Some well-known techniques that come at the

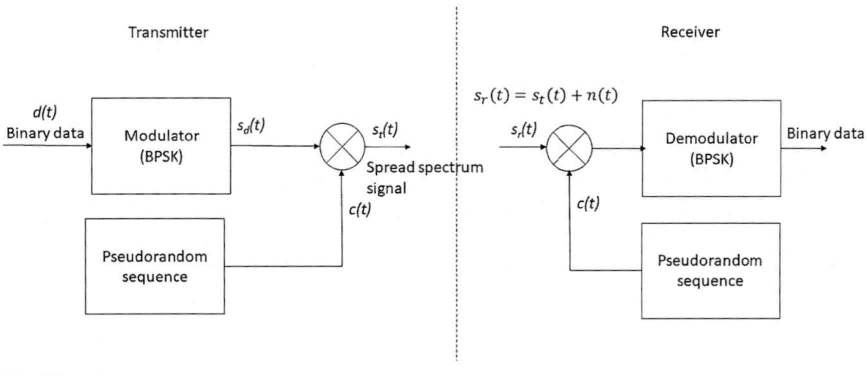

FIGURE 6.11

Direct sequence spread spectrum transmitter and receiver.

cost of increased computation have been discussed in Ref. [15]. The work done in Ref. [6] addresses this problem much more efficiently by using a spread spectrum technique to scramble the transmitted data. More specifically, it uses a technique known as DSSS, which is broadly utilized in IEEE 802.11 standards to make the signal resistant to interference [31]. DSSS uses a special type of pseudorandom sequence (one that is random in nature but repeats itself after a predefined interval) to modify the bits to be transferred. For example, a PN sequence of order 9 will repeat after 2^9-1 bits.

$$s_d(t) = A \times d(t) \times \cos(2\pi f_c t) \tag{6.4}$$

$$s_t(t) = A \times d(t) \times c(t) \times \cos(2\pi f_c t) \tag{6.5}$$

$$s_t(t) \times c(t) = A \times d(t) \times c(t) \times \cos(2\pi f_c t) \tag{6.6}$$

$$s_t(t) \times c(t) = s_d(t) \tag{6.7}$$

A simplified version of the DSSS transmitter and receiver is shown in Fig. 6.11. The input signal $s_d(t)$ is defined as Eq. (6.4). By multiplying $s_d(t)$ with a pseudorandom sequence $c(t)$, it is converted into a noise-like wide bandwidth signal. By doing so, each bit in the original signal is represented by multiple bits. For example, a 10-bit spreading code spreads signal such that it occupies 10 times more space in the frequency domain. The transmitted signal $s_t(t)$ is represented by Eq. (6.5). The receiver circuit multiplies the received signal $s_r(t)$ with the pseudorandom sequence, as shown in Eq. (6.6). As the convolution of the pseudorandom sequence by itself results in a unity function, the output of the convolution process results in the original signal at the receiver, as shown in Eq. (6.7). Note that this simple model does not take noise into account. Essentially, DSSS is an encryption procedure to conceal the information by pushing it under the noise floor of the signal. It is difficult to extract information from the noise-like-signal sequence because the signal energy is spread over a larger bandwidth. The only way this can be done is if the receiver has access to the same key (pseudorandom sequence) that was used at the transmitter. M-sequence and Gold-sequence are some of the well-developed techniques to generate a pseudorandom sequence. The secret key, which is the seed value used for the linear feedback system, can help aid in determining the output sequence. Generally, the two communicating parties have agreed upon a preshared secret key to generate the same pseudorandom sequence and then share this information in a secretive way with each other. Advantages, such as resistance to jamming and hard to demodulate, make DSSS a good way to prevent eavesdropping.

The work presented in this section is significant as it can protect the IMDs from a jamming attack at the physical layer, wherein the frequency band being used by the IMD can be jammed by other high-power signals. Similarly, tampering attacks in which contents of the signal are modified can also be avoided by the spread spectrum sequence technique described in this section. By preventing the jamming attacks as described earlier, the technique presented here also avoids inefficient network performance. Such attacks, known as resource exhaustion, happen when the intruder tries to flood the network with repeated messages to generate collisions. In summary, the work presented here can significantly improve the security of the IMDs by protecting the communication layer against attacks at the physical layer.

7. Application layer security: in-device intrusion detection

In terms of in-device intrusion detection, this section focusses on the classification problem in the context of an insulin pump. There are a diverse set of security aspects namely authentication and forecasting. The authentication problem for cardiac defibrillator and forecasting in the context of deep brain stimulators is explained in Refs. [6,10], respectively.

7.1 Insulin pump therapy

An insulin pump is a small external device that delivers quick insulin 24 hours every day. With most frameworks, the pump is attached to a plastic tube that has a plastic needle that is under the skin, for the most part of the stomach area. The infusion set, that is, the tube and needle must be changed each 2–3 days as indicated by every producer's guidelines.

For years now, patients have used insulin pumps to control the sugar level in their blood. Traditionally, this has been done by using syringes and manually determining the amount of insulin to be used. Today, this can be done using a wireless insulin pump (WIP), whose framework encompasses a blood glucose monitor, a perpetual glucose management system, and an insulin pump with other related gadgets, all connected via wireless links as shown in Fig. 6.12. It considers greater granular insulin transmission while attaining blood glucose control. Although this is a big boon for patients, there are many security risks that come in parallel to this multifaceted nature of technological marvel.

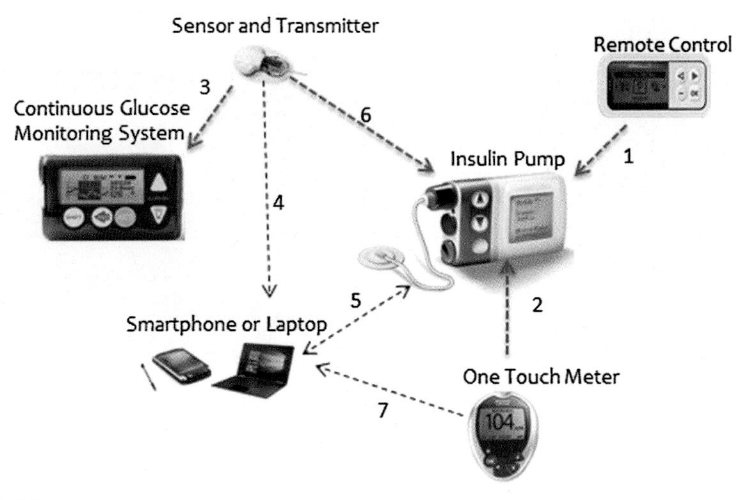

FIGURE 6.12

Insulin pump therapy.

Reproduced from Rathore, H., Al-Ali, A., Mohamed, A., Du, X. and Guizani, M., December 2017. DLRT: deep learning approach for reliable diabetic treatment. In GLOBECOM 2017-2017 IEEE Global communications conference. pp. 1–6. IEEE.

Machine learning is used to distinguish between the genuine and false dosage of glucose. The performance of the specific technique is ascertained using data samples found in Ref. [29]. To assess the performance of the proposed methodology, 70 representative datasets, with each set having 1000 samples on average, were used to run the experiments. A typical WIP takes every data point into consideration (point by point data), which includes patient information such as the measurements related to infusion rate and associated time-stamps. An output label is provided on the basis of code deciphered based on inputs such as insulin and glucose intake. For example, regular insulin dosage has a code 33 (output label as 0), whose value relates to.

A multilayer perceptron (MLP) neural network was designed with an input layer, output layer, and three hidden layers (h). The input dimension was set on five and the output dimension on one. The hidden layers contain four, three, and two neurons, respectively. The model initiation parameters were setup with batch size as 10 and epoch as 150. Rectifier activation function was used at the initial layers, and the sigmoid activation function was used at the output layer. Adam optimizer was used to specify the loss function for assessing the weights [30]. Simulations were carried out on a larger dataset of 30,000 samples with the validation split of 33%, hidden layers (h) = 3, 93.98% mean accuracy was achieved. 93.98% average accuracy with a standard deviation of 4.92% was estimated by k-fold cross-validation. The results inculcate that deep learning has higher accuracy, precision, recall, and f1-score with respect to support vector machine (SVM).

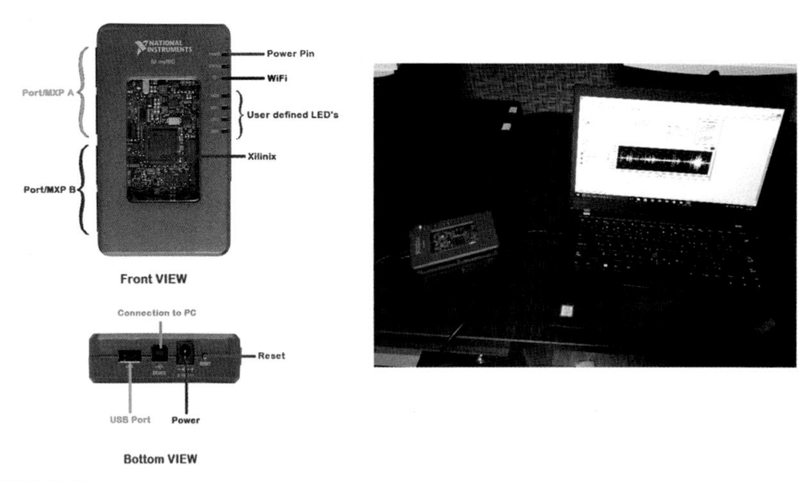

FIGURE 6.13

(A) NI myRIO diagram. (B) Picture of the testbed.

The software simulations were further validated on field-programmable gate array (FPGA). For experimental analysis, NI myRIO real-time embedded system was used as depicted in Fig. 6.13. This device was programmed using NI LabVIEW 2015 and NI Real-time module, along with the associated drivers. A cloud-based compile server was used to offload FPGA compilations. This prototyping board includes analog and digital input/output lines, sensors such as onboard accelerometers and numerous controls and indicators. For processing, the board includes Xilinx Virtex7 FPGA with a dual-core ARM Cortex-A9 processor.

A programmatic FPGA interface was used to manage the communication between the FPGA and the host. It should be noted that the function that currently runs on the FPGA (used a prototyping platform) will eventually run on the system (at the insulin pump) and the host VI runs on the host computer. The host computer makes decisions on the amount of insulin and collects readings from the insulin pump. The FPGA runs in timed loops at 25 or 40 MHz and is responsible for initiating the activation function, data processing, and staging processing. The selection of the activation function and informing the FPGA of the decisions made is done by the host. NI's myRIO is a rapid prototyping platform that allows us to assess the performance of different algorithms and strategies for resource utilization. Once the prototyping is finished, the same algorithm can be moved to a real-world device with less room for implementation-related suspense.

We compared multilayer perceptron model with SVM on FGPA on matrices such as number of slices, look-up table (LUT), slice register, block RAM, and DSP48 as given in Table 6.1. Number of slices on an FPGA is a metric of the hardware resources that the FPGA has. A slice consists of LUTs and flip flops. The number of LUTs and flip flops that Xilinx defines to make up a single slice is different based on the family of the chip. A LUT is a small asynchronous random-access memory

Table 6.1 Comparison of final device utilization (placement) for multilayer perceptron model versus support vector machine.

Device utilization	MLP (Percent)	SVM (Percent)
Total slices	100.0	78.2
Slice registers	52.8	28.8
Slice LUTs	87.8	52.3
Block RAMs	30.0	8.3
DSP48s	38.8	5.0

Table 6.2 Comparison of multilayer perceptron model with support vector machine.

	MLP	Linear SVM
Accuracy	98.1%	90.17%
Precision	98.16%	88%
Recall	99.83%	90%
F1-score	98.98%	86%
Time complexity	$O(n^{h+1}*s)$	$O(ns^3)$

Note: n *is the number of features,* s *is the number of samples,* h *is the number of hidden layers where* n = *10,* s = *12,933,* h = *2. MLP has 100*12,933 computations to be performed, which is* $<<10*(129,33^3)$ *in case of linear SVM.*

that can be used to implement combinational logic. It can also be used to store calibration coefficients or filter coefficients. A slice register is like a flip-flop, which is a single-bit memory cell used to hold state. Block RAMs can be used to hold small chunks of data. The size of the data is based on the RAM size on the FPGA. This can be used to hold historical data, such as training data for machine-learning algorithms. DSP48 is a floating-point processor on the FPGA used for doing floating-point mathematics on the FPGA.

In addition, we used accuracy, precision, recall, f1-score, and time complexity as metrics, in Table 6.2, to compare the proposed MLP model with linear SVM.

To conclude, MLP achieves better accuracy, precision, and recall, but it comes at a slightly higher resource utilization on the FPGA as compared to the linear SVM.

8. Closing remarks

IMD is a type of cyberphysical system that uses sensing, communication, and control capabilities to significantly improve the quality of life for patients. Needless to say, this technology has been gathering a lot of excitement with medical field professionals around the world. Such devices closely mimic the human brain in their functionality. Hence, it makes sense that one can learn from how the human brain protects the body and apply this learning to securing cyberphysical systems as well. This chapter describes the work that we have done in this area. It uses deep learning, a subset of machine learning that learns from the inherent patterns in the data for solving a diverse set of problems such as recognition, classification, and segmentation. It provides security at the sensor, communication, and application layers using novel techniques. This is critical as IMDs now have 802.11x or LTE chips on with the goal that they can converse with one another, in addition to the conventional jobs of sensing and actuating.

Although deep learning is a very powerful tool for AI, as exhibited by various models in the chapter, a word of advice and motivation for more innovation in this area is well justified at this juncture. For us as human beings, it takes a long time to develop experiences, reasoning, and emotional maturity and it takes years to master. It is hard to quantify how much training data are involved in this endeavor. The state-of-the-art machine-learning algorithms, at their best, mimic the cerebrum of a baby. Psychologists refer to this stage in the baby's growth as a sponge wipe, which soaks in whatever comes its way and learns through preparations. Deep-learning techniques have a long way to go before making up for evolution learnings with the ages of advancement that the human mind has experienced. For example, in the realm of image processing, today's algorithms, at the best, can resemble the sparsity achieved by a couple of layers of neurons on the external part of the retina, which does an introductory job of preparing a picture. It is very unlikely that such a network could be bent to all the tasks our brains are capable of. As these systems do not know things about the world the way a genuinely clever person does, they are fragile and effectively confounded.

The second thing to remember is that a human being's individual's comprehension of things is unmistakably increasingly unrivaled. As a child, our guardians tell us that lions are dangerous and by the method of induction, we infer that bears can be dangerous too. We do so by associated similar patterns (such as sharp teeth, growling sound, size) between the two animals. This is an artifact of deep reasoning found in humans, not currently seen in deep learning. Deep learning does utilize derivation but does so in a straight, fundamental, and one-dimensional way. If the network is trained to classify lions as dangerous, then it is most likely going to be sensitive only to that species. It would not automatically associate this training to a bear. As a related example, an AI trained machine would not deduce that a leopard belongs to the cat family, even though it is trained to identify a cat.

Scientists envision a future where machines can not only sense and make a decision, but can also learn from thoughts and emotions. This would expect researchers to think of different sorts of artificial neural networks, and in some cases giving them natural, premodified learning—like the impulses that every single living thing is brought into the world with. However, until this occurs, we have to hand-code into them a great deal of existing human learning. Level 5 autonomy in self-driving cars is going through this phase where these vehicles are being driven for miles at a time so that they can learn from road conditions. They are essentially trying to mimic what human drivers have done for ages. Although we start driving at the age of 16 or 18, the learning process starts with backseat driving from a very young age and learning from others driving around us.

References

[1] Carey J. Brain facts: a primer on the brain and nervous system. 1990. 1990.

[2] Kramer DB, Baker M, Ransford B, Molina-Markham A, Stewart Q, Fu K, Reynolds MR. Security and privacy qualities of medical devices: an analysis of FDA postmarket surveillance. PLoS One 2012;7(7):e40200. 2012.

[3] Pycroft L, Boccard SG, Owen SL, Stein JF, Fitzgerald JJ, Green AL, Aziz TZ. Brainjacking: implant security issues in invasive neuromodulation. World neurosurgery 2016; 92:454−62. 2016.

[4] Zion market research, https://www.globenewswire.com/news-release/2018/09/21/1574327/0/en/Global-Active-Implantable-Medical-Devices-Market-Worth-Over-USD-23-33-Billion-by-2024-Zion-Market-Research.html [accessed on 10 June, 2019].

[5] Rathore H, Al-Ali A, Mohamed A, Du X, Guizani M. DLRT: deep learning approach for reliable diabetic treatment. In: GLOBECOM 2017-2017 IEEE Global communications conference. IEEE; December 2017. p. 1−6. 2017.

[6] Rathore H, Fu C, Mohamed A, Al-Ali A, Du X, Guizani M, Yu Z. Multi-layer security scheme for implantable medical devices. Neural Computing & Applications 2018: 1−14.

[7] Rathore H, Mohamed A, Al-Ali A, Du X, Guizani M. A review of security challenges, attacks and resolutions for wireless medical devices. In: 2017 13th International wireless communications and mobile computing conference (IWCMC). IEEE; 2017. p. 1495−501.

[8] Rathore H, Wenzel L, Al-Ali AK, Mohamed A, Du X, Guizani M. Multi-layer perceptron model on chip for secure diabetic treatment. IEEE Access 2018;6:44718−30.

[9] Rathore H, Al-Ali A, Mohamed A, Du X, Guizani M. DTW based authentication for wireless medical device security. In: 2018 14th International wireless communications and mobile computing conference (IWCMC). IEEE; June 2018. p. 476−81. 2018.

[10] Rathore H, Al-Ali A, Mohamed A, Du X, Guizani M. A novel deep learning strategy for classifying different attack patterns for deep brain implants. IEEE Access 2019.

[11] Leveson NG, Turner CS. An investigation of the Therac-25 accidents. Computer 1993; 26(7):18−41.

[12] Hauser RG, Maron BJ. Lessons from the failure and recall of an implantable cardioverter-defibrillator. Circulation 2005;112(13):2040−2.

[13] Hei X, Du X. Biometric-based two-level secure access control for implantable medical devices during emergencies. In: INFOCOM, 2011 Proceedings IEEE. IEEE; 2011. p. 346−50.

[14] Halperin D, Heydt-Benjamin TS, Ransford B, Clark SS, Defend B, Morgan W, Fu K, Kohno T, Maisel WH. Pacemakers and implantable cardiac defibrillators: software radio attacks and zero-power defenses. In: Security and privacy, 2008. SP 2008. IEEE Symposium on. IEEE; 2008. p. 129−42.

[15] Rostami M, Juels A, Koushanfar F. Heart-to-heart (H2H): authentication for implanted medical devices. In: Proceedings of the 2013 ACM SIGSAC conference on computer and communications security. ACM; 2013. p. 1099−112.

[16] Kim B, Yu J, Kim H. In-vivo nfc: remote monitoring of implanted medical devices with improved privacy. In: Proceedings of the 10th ACM conference on embedded network sensor systems. ACM; 2012. p. 327−8.

[17] Zheng G, Fang G, Shankaran R, Orgun MA, Dutkiewicz E. An ECG-based secret data sharing scheme supporting emergency treatment of implantable medical devices". In: Wireless personal Multimedia communications (WPMC), 2014 International Symposium on. IEEE; 2014. p. 624–8.
[18] Ankaralı ZE, Demir AF, Qaraqe M, Abbasi QH, Serpedin E, Arslan H, Gitlin RD. Physical layer security for wireless implantable medical devices. In: Computer aided Modelling and Design of communication links and networks (CAMAD), IEEE 20th International Workshop on. IEEE; 2015. p. 144–7.
[19] Zhang M, Raghunathan A, Jha NK. MedMon: securing medical devices through wireless monitoring and anomaly detection. IEEE Transactions on Biomedical circuits and Systems 2013;7(6):871–81.
[20] Gupta S. Implantable medical devices-cyber risks and mitigation approaches. In: Proceedings of the Cybersecurity in cyber-physical Workshop. US: The National Institute of Standards and Technology (NIST); 2012.
[21] Rieback MR, Crispo B, Tanenbaum AS. RFID Guardian: a battery-powered mobile device for RFID privacy management. In: Australasian conference on information security and privacy. Springer Berlin Heidelberg; 2005. p. 184–94.
[22] Hei X, Du X, Wu J, Hu F. Defending resource depletion attacks on implantable medical devices. In: Global Telecommunications conference (GLOBECOM 2010). IEEE; 2010. p. 1–5. IEEE, 2010.
[23] Singh K, Muthukkumarasamy V. Authenticated key establishment protocols for a home health care system. In: Intelligent sensors, sensor networks and information, 2007. ISSNIP 2007. 3rd International conference on. IEEE; 2007. p. 353–8.
[24] Denning T, Fu K, Kohno T. Absence makes the heart grow fonder: new directions for implantable medical device security'. In: HotSec; 2008.
[25] Xu F, Qin Z, Tan CC, Wang B, Li Q. IMDGuard: securing implantable medical devices with the external wearable guardian". In: INFOCOM, 2011 Proceedings IEEE. IEEE; 2011. p. 1862–70.
[26] Gollakota S, Hassanieh H, Ransford B, Katabi D, Fu K. They can hear your heartbeats: non-invasive security for implantable medical devices. ACM SIGCOMM – Computer Communication Review 2011;41(4):2–13.
[27] Rasmussen KB, Castelluccia C, Heydt-Benjamin TS, Capkun S. Proximity-based access control for implantable medical devices. In: Proceedings of the 16th ACM conference on Computer and communications security. ACM; 2009. p. 410–9.
[28] Xu F, Qin Z, Tan CC, Wang B, Li Q. IMDGuard: securing implantable medical devices with the external wearable guardian. In: Proc. IEEE INFOCOM; 2011. p. 1862–70.
[29] Lichman M. UCI machine learning repository. Irvine, CA: University of California, School of Information and Computer Science; 2013. http://archive.ics.uci.edu/ml.
[30] Kingma D, Ba J. Adam: a method for stochastic optimization. arXiv preprint arXiv: 1412.6980. 2014.
[31] Rappaport TS. New Jersey: prentice hall PTR. Wireless communications: principles and practice, vol. 2; 1996.

CHAPTER

Secure medical treatment with deep learning on embedded board

7

Abderrazak Abdaoui[1], Abdulla Al-Ali[1], Ali Riahi[1], Amr Mohamed, PhD[3], Xiaojiang Du[2], Mohsen Guizani[3]

[1]*Department of Computer Science and Engineering, Qatar University, Doha, Qatar;* [2]*Department of Computer and Information Sciences, Temple University, Philadelphia, PA, United States;* [3]*Professor, Department of Computer Science and Engineering, Qatar University, Doha, Qatar*

1. Introduction

Deep brain stimulation (DBS) is an excellent clinical method for movement disorders that no longer responds satisfactorily to pharmacological management, but its progress has been hampered by stagnation in technological procedure solutions and embedded device development [1]. Nowadays, DBS is the sole method to treat patients with a variety of movement disorders. This method is based on the implantation of one or more electrodes in the brain to interrupt and stimulate nerve activities [2]. Other disorders can be treated by DBS include essential tremor and dystonia such as genetic dystonia, generalized dystonia, hemidystonia, and segmental dystonia [3].

An implantable pulse generator (IPG), implanted in the chest region, is connected to the deep brain stimulator and continuously leads its functionality. The IPG contains a device source that generates the electrical pulses and one battery for both the DBS and the IPG [4]. The patient or the clinician can turn on or off the IPG via a small remote control. In addition, for the electrical stimulation, the clinician can select the desired electrodes on each brain to be activated. This gives the flexibility of activating the stimulation to a particular part of the brain. The electrical parameters of the pulse generator can be adjusted by the DBS programmer such as a doctor or nurse practitioner. The adjustments allow the clinician to regulate the DBS system to maximize the benefits and minimize the side effects. The stimulation signals, produced by the device, can be personalized by programming and reprogramming the IPG for symptoms and symptom changes [5]; the procedure is reversible and it can be performed safely on both sides of the brain.

In Swann et al., [6], the authors introduced an adaptive DBS for two patients having Parkinson's disease by an implanted neural prosthesis that is enabled to take brain sensing to control the stimulation amplitude. They used a cortical gamma oscillation (60–90 Hz) to reduce stimulation voltage when gamma oscillatory activity is high and increase stimulation voltage when it is low. In their approach, the

Energy Efficiency of Medical Devices and Healthcare Applications. https://doi.org/10.1016/B978-0-12-819045-6.00007-8

authors, demonstrate the importance of "adaptive deep brain stimulation" by performing stimulation on two persons with Parkinson's disease. In Ref. [7], the authors presented a closed-loop DBS platform for investigating control strategies for the management of essential tremor. They demonstrated a system capable of using a variety of sensors for the inertia, electromyography, and neurostimulator electrode readings. The stimulation is adapted to the sensed data. Their solution could result in lower average power dissipation and reduced side effects from unneeded stimulation.

These innovative approaches of DBS are joined with a huge security risk. Indeed, an adversary can take control of the implantable pulse generator to produce fake signals to damage the brain of the patient [8–13]. Although the advantages offered by DBS are substantial, there is an increasing risk of devices to be manipulated remotely by an attacker (hacker), disabled or subverted because of weak cybersecurity. Hackers can benefit from the wireless control features to manipulate the DBS settings, while networked DBS are vulnerable for attacks from any hackers in the world. The rapid huge increase in the number of DBS and the increasing variety of features is very risky and alarming. It depends on the role that the patient plays in the society (i.e., politician, rich person, etc.), the vulnerable DBS will motivate the assassination of these patients.

In Ref. [14], the authors give an overview of the main security goals for the next generation of implantable medical devices and analyze the most relevant protection mechanisms proposed so far. First, the security proposals must give more attention to the inherent constraints of these small and implanted devices: energy, storage, and computing power. Second, the proposed solutions must achieve an adequate tradeoff between the safety of the patient and the complexity of deployment [15].

The DBS frameworks include specific electrode embedded into the human cerebrum, inline expansions either running behind or embedded inside the ear, and an implantable pulse generator embedded either on the top of or inside the region above the chest as shown in Fig. 7.1. In this chapter, we design and we introduce a complete prototype of an embedded system to predict different attack patterns in DBS. We implemented the deep-learning methodology in the embedded device. We proved the robustness of the proposed device by emulating several random attacks on the stimulator. Results show that our system is 97% reliable to predict attacks. Our contribution focus on the following points:

- Design the deep classifier to predict the genuine signal or the attack within a large number of attack patterns existing till now.
- Implement the whole solution on a web application running on an embedded device, such as Raspberry Pi3 to ease its integration on an existing private network of the hospital.
- The server supporting the deep classifier (main attack defender) could be available for several patients and signal emulators.

The remainder of the chapter is organized as follows. Section 2 presents the related work for Implantable Medical Device (IMD) and DBS security. Section 3 pictures the deep neural network classifier for DBS. Section 4 describes the implementation of deep neural network classifier on Raspberry Pi3. Section 5 presents the

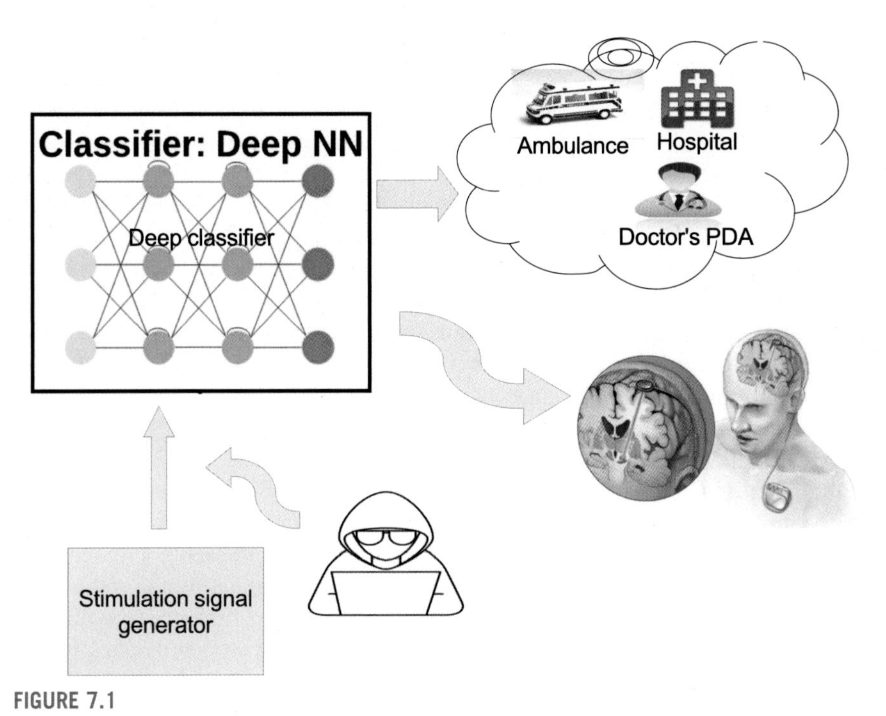

FIGURE 7.1
Deep classifier in DBS.

results related to the accuracy of the learning step and the performance evaluation of the deep neural network classifier when the signals and the attacks are emulated. Section 6 concludes the chapter.

2. Related works

Security of the communication, in implantable medical devices (IMDs), is among the most critical issues for the safety of the patient, and several research groups are focusing their efforts on this important subject. However, designing a reliable solution guaranteeing the security of IMD is conflicting with several issues related to the basis of the IMD itself and the environment surrounding the patient. These issues are dependent on, and not limited to, the battery life of the device, the adaptability, availability, and the requested reliability of the secure solution. In this section, we present an overview of the previous works related to the security of the communication links of the implantable medical devices in general and the DBS particularly.

In Ref. [16], Heena et al. presented an on-chip neural network system for the security of devices involving insulin pumps used by diabetic patients. They showed that their system achieved 98.1% accuracy in classifying fake versus genuine glucose measurements.

In another work [17], Heena et al. proposed a highly accurate and efficient deep-learning methodology to protect vulnerable devices against fake glucose dosage. They proved that their proposed method outperforms the state-of-the-art accuracy to achieve 93% and 90% when all the wireless links are secured.

In another work, the authors in Ref. [18] presented a review of various attacks and strategies used in overcoming the attacks in IMD. They focused their efforts on the challenges, the threats, and the solutions pertaining to the privacy and the safety issues of medical devices. Adversaries surrounding an IMD user may threat the patient in different ways. Usually, this adversary aims to sense private health information sent by IMD or actively attempts to modify IMD parameters. In addition, in a worst case, a group of adversaries may be located nearby the patient, coordinate to catch information from the unit controlling the IMD. Various classifications of IMD security issues are provided in the literature [19–22].

IMD can be hacked and fatal operations may be performed, such as ordering an insulin pump to apply an overdose insulin injection [12,23] or an implantable cardiac defibrillator to emit a chock designed to reduce a fatal heart rhythm [24].

In Ref. [25], the authors presented an independent approach for monitoring poisoning attacks across a wide range of machine-learning algorithms and healthcare datasets. In their proposed scheme, the authors have experimented, six different machine-learning algorithms. Moreover, they show that the proposed attack is successful even without prior knowledge of the machine-learning algorithm details. Their work is limited to the tests of algorithms on a normal computer without going to the pure implementation of the prediction.

The authors in Ref. [26] proposed a classification of the attacks in DBS in two types: blind attacks and targeted attacks. Blind attacks include the cessation of stimulation, draining implant batteries including tissue damage, and information theft. However, targeted attacks include impairment of motor function, alteration of impulse control, modification of emotions, and induction of pain.

In this chapter, we would investigate these attacks to improve the security of the DBS by adding an intermediate component to classify the stimulation signal into fake or genuine before being sent to the electrodes of the DBS.

In Ref. [9], the authors proposed a robust approach for securing against existing and potential communication-based attacks on IMDs while keeping the added hardware and power consumption low. In addition, a new efficient and secure protocol for authorizing third party medical teams to access the IMDs in the case of emergency was introduced. However, their contribution is limited to the presentation of the authentication protocol without any hardware implementation of prototypes.

3. Deep neural network classifier for deep brain stimulation

3.1 Network and attack model

The goal of DBS is to superimpose a stimulation pattern over the patient's chronic pain pattern and to establish a correct stimulation waveform with the help of amplitude, the pulse width of the stimulation signal.

When an external reader tries to connect to the IMD, authentication followed by the communication with the IMD is performed. Once the adversary succeeds to communicate with the IMD, it can modify the stimulation patterns, to introduce an acute stimulation pattern inside the brain. Different types of parameters can be introduced to disrupt the normal behavior of the DBS [26]. In this chapter, we propose and implement a deep-learning methodology to predict different attack patterns in deep brain stimulators. The system consists of data source(s), for example, (1) pulse generator for brain stimulation, (2) server node: raspberry pi, and (3) implanted medical device.

The data source, known as a client node, is responsible for the emulation/generation of the stimulation signals. The server node is responsible for data processing: classification of the stimulation signal to predict if it is fake or genuine measurement.

3.2 Machine-learning basis

Deep learning, *or deep structured learning*, is a new area of machine learning based on artificial neural networks (ANN), which is bioinspired from human brain neurons [27]. Artificial neural networks were inspired by information processing and distributed communication nodes in biological systems [28]. Compared to existing system models, which need regular updates and replacement, ANN models maintain themselves and even learn during changing conditions. Each node of the ANN is just a mathematical function. Deep learning is one of the many machine learning to enable a computer to perform tasks such as stock prediction, house rent prediction, and image and signals classification. A neural network, in general, can be divided into two types: convolutional neural network and artificial neural network. A convolutional neural network is commonly applied for image processing like the classification of images into various groups, the presence of a tumor, etc.

(1) *Sequence neural network model:* In real-world scenarios, the sequence model is employed in speech recognition, sentiment classification, DNA sequence analysis, machine translation, and signal sequence classification. In our system, we considered a sequence neural network model with L layers, containing $L-2$ hidden layers.

(2) *Features extraction:* This step of prelearning consists of transforming the raw data into meaningful inputs for the machine-learning block. Feature extraction was first designed through a handcrafted process by experts in the field [29]. This step starts from an initial set of measured data and builds derived values called features assumed to be informative and nonredundant to make ease of subsequent learning. The extracted features are assumed to contain the relevant information from the raw data, so that the desired task can be realized by employing this reduced representation, instead of the complete raw data [30]. Features are always learned directly from the data, similar to the weights in the classifier. The learned features are used in a machine learning form called deep neural networks (DNNs).

(3) *Classification:* Feature classification is a pattern recognition technique that is used to categorize a huge number of data into different classes based on some criteria. Sometimes feature classification can also be related to feature selection, which consists of selecting a subset of the extracted features that would optimize the machine-learning algorithm and possible reduce noise removing unrelated features [31].

3.3 Deep neural network design

For the design of the deep learning model, we consider a sequence neural network model with L layers containing $L-2$ hidden layers and one input layer and one output layer. We denote by X the elements of the input sequence generated by the signal stimulator.

$$X=(x_1,\ldots,x_L)\in E_x\times,\ldots,\times,E_x=E_x^L, \tag{7.1}$$

we can also write the Recurrent Neural Networks (RNN) target variables, denoted by y as a sequence of bounded length:

$$Y=(y_1,\ldots,y_L)\in E_y\times,\ldots,\times,E_y=E_y^L, \tag{7.2}$$

where E_x is an inner product space. The datasets will be of the form $\mathcal{D}_L=\{(x(i),\ y(i))\}_{j=1}^{n}$; however, the sequences are generally of variables length, and we assume that the input layer has 512 nodes, the first hidden layer contains 256 nodes with activation function "sigmoid" defined as follows:

$$\psi_i(x)=\frac{1}{1+\exp(-x)}, \tag{7.3}$$

the second hidden layer is composed of 128 nodes with activation function "ReLU" defined as follows:

$$\psi_i(x)=\max(0,x) \tag{7.4}$$

and the output layer with `num-classes` nodes and activation function "softmax" defined as follows:

$$\psi_i(x)=\frac{\exp(x)}{1+\exp(x)} \tag{7.5}$$

In our deep classifier, the target Y defines the labels related to the sequence status. To set up the values of these labels, we consider the two following scenarios.

Deep neural network classifier: (attacked)/(nonattacked). In this scenario, our proposed system is able to predict, in the presence of several kinds of attacks, whether the signal contains an attack, or genuine, without giving any information about the kind of attack. In this case, the labels are defined as follows:

$$y_i\in\{0:\ (\textit{Genuine signal});1:\textit{Attack}\} \tag{7.6}$$

where *Attack* could be one of the following: $Attack_1, \ldots, Attack_{n_a}$, `num_class = 2`, where n_a is the number of attack strategies.

Deep neural network classifier: got specific attack/nonattacked: The deep neural network predictor is able to classify the entry x, by returning a precise information on the attack pattern:

0	1:	2:	3:	4:	5:	6:	7:
	$Attack_1$	$Attack_2$	$Attack_3$	$Attack_4$	$Attack_5$	$Attack_6$	$Attack_7$
Genuine: nonattacked	Spike	Outlier	Stuck	Incremental	Chronic	Noise	Unusual

$$y_i \in \{0: \textit{Genuine signal}; 1: Attack_1; \ldots; n_a: Attack_{n_a}\} \tag{7.7}$$

`num_class` $= n_a + 1$, here y corresponds to the class of the data sequence, (stimulation signal transmitted to the deep brain). We constraint the output of the DNN to be a valid discrete probability distribution. This can be done by applying the "softmax" function $\zeta(x)$ to the network output $\zeta(x, \theta)$. The prediction of the label is then given by

$$\widehat{y}_L = \zeta(\mathcal{F}(x, \theta)) \tag{7.8}$$

1) *Optimization method:* Most optimization methods are gradient based, this means that we must calculate the gradient $\mathcal{J}$ with respect to the parameters at each layer $i \in \{1, \ldots, L\}$. In the neural network, the optimization method should produce slightly better and faster results by updating the model weights and the bias values [32]. These methods may be as follows:

- Gradient descent,
- Stochastic gradient descent,
- Adam,

Gradient descent: This optimization method is one of the most popular algorithms to optimize neural networks. This method minimizes an objective function $\mathcal{J}$ parametrized by the model parameter θ by updating in the backward direction of the objective function $\nabla_\theta(\mathcal{J}(\theta))$ with respect to the parameters:

$$\theta_{t+1} = \theta_t - \alpha \nabla_\theta(\mathcal{J}(\theta)) \tag{7.9}$$

Stochastic gradient descent: It is an iterative optimization function with a specific smoothness property. Stochastic gradient descent uses random samples to evaluate the gradient:

$$\theta_{t+1} = \theta_t - \alpha \nabla_\theta(\mathcal{J}(\theta, \gamma, \kappa)), \tag{7.10}$$

where γ, κ are a few training samples called minibatch, statistics taken from the previous training samples.

Adam: This method works well in practice and outperforms all other adaptive techniques as it converges very fast. With adam, the problems of the vanishing learning rate, slow convergence, or high variance in the parameter updates are resolved [28]. Regarding its robustness in practice, in our deep-learning system, we adopted the optimization based on "adam" method.

(2) *The evaluation techniques in machine learning:* For the evaluation of machine learning, the metrics are very important. Indeed, the choice of metrics influences how the performance of machine learning is measured and compared.

Classification metrics: Classification accuracy is defined by the number of correct predictions made as a ratio of all predictions made. This is the most common evaluation metric for classification problems. For our specific application of security of the communication in implantable medical devices, we shall use a more specific metric to take into account the false detection of an alarm and the nondetection of an alarm. We consider the two following events:

- True positive (*TP*), the event that the classifier makes prediction of an attack and in fact there is an attack.
- False positive (*FP*), the event when the classifier makes prediction of an attack and in fact there is no attack.

From these events, we got the two metrics: positive predictive value (PPV) and negative prediction value (NPV).

$$\begin{aligned} \text{PPV} &= \frac{\#\text{ of true positive}}{\#\text{ of true positive} + \#\text{ of false positive}} \\ &= \frac{TP}{TP + FP}. \end{aligned} \tag{7.11}$$

However, the negative predictive value is defined as follows:

$$\begin{aligned} \text{NPV} &= \frac{\#\text{ of true negative}}{\#\text{ of true negative} + \#\text{ of false negative}} \\ &= \frac{TN}{TN + FN}. \end{aligned} \tag{7.12}$$

where true negative (*TN*) is the event that the classifier makes prediction of a genuine signal and in fact the signal is genuine. However, false negative (*FN*) is the event when the classifier makes prediction of genuine signal and in fact the signal has an attack.

The sensitivity is defined as follows:

$$\text{S} = \frac{TP}{TP + FN} \tag{7.13}$$

The specificity is defined as follows:

$$\text{Specificity} = \frac{TN}{FP + TN} \tag{7.14}$$

$$\text{Accuracy} = \frac{\#\text{correctly classified items}}{\#\text{all classifed items}} \tag{7.15}$$

Logarithmic loss is used to evaluate the predictions of the probabilities of membership to a given class. $\alpha \in 2\ [0\ 1]$ can be seen as a measure of the confidence for a prediction. In our work, we consider the evaluation based on accuracy.

The implementation of deep learning is summarized in the following part of the code:

```
import keras
from keras.models import Sequential
from keras.layers import Dense
from keras.preprocessing import sequence
from sklearn.model_selection import train_test_split
# create the model
embedding_vecor_length = 32
model = Sequential()
model.add(Dense(512, activation = 'sigmoid'))
model.add(Dense(256, activation = 'sigmoid'))
model.add(Dense(128, activation = 'relu'))
model.add(Dense(num_classes, activation = 'softmax'))
model.compile(loss = 'binary_crossentropy', optimizer
 = 'adam', metrics = ['accuracy'])
model.build()
history = model.fit(X_trains, y_trains, epochs = Epochs,
batch_size = 64,validation_data=(X_tests, y_tests),
verbose = 2, shuffle = False)
# Final evaluation of the model
scores = model.evaluate(X_tests, y_tests, verbose = 0)
#final step of the learning, save the complete NN model after learning the whole
parameters using the dataset. model.save('my_modelXX.h5')
```

The part of the python code used for the learning step, the complete neural network model with all the updated coefficients, is finally saved in HDF5 format. This model will be loaded and employed in the python code dedicated to the prediction or the deployment of the classifier on the embedded board (server). Note that the learning step usually consumes excessive CPU and memory resources. Its implementation requires a powerful PC, or in the worst case, a PC with GPU. CPUs are suited for more general computing workloads. However, GPUs are designed to compute in parallel the same instructions. Deep neural networks are structured in a very uniform manner such that at each layer of the network thousands of identical artificial neurons perform the same computation. In our implementation, we use a simple PC 8 GB memory, 3.6 GHz, with Linux (ubuntu 16.04). The flowchart of the learning step is described in the flowchart, Fig. 7.2.

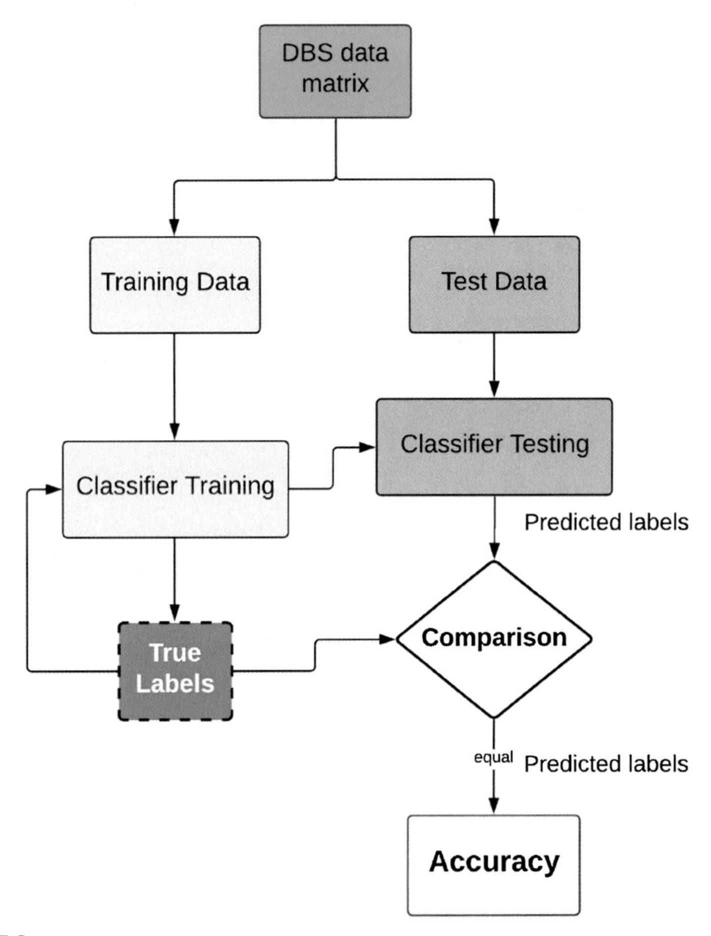

FIGURE 7.2

Flowchart of the learning step.

4. Hardware implementation

The solution for the secure DBS is composed of one client generating stimulation signals and attack patterns and one server receiving requests from the client via TCP/IP socket link. Both of them are supporting web application to run the micro-web framework Flask and the python modules for data processing [33]. In the existent solution applied nowadays, the signals generated by the implanted pulse generator are vulnerable to external attacks and may cause dangerous troubles to the patient or in worst case death. The new element we add to the DBS solution ensures the accurate security of the stimulation signals and the safety of the patient.

4.1 Web applications design

On the client and on the server, a web application is running to support the whole application on each side. During the design of each web application, we employed a dashboard, which is monitored and written in Python using flux advanced security kernel (Flask). Flask is a microframework, which does not need specific tools or libraries [33]. Flask supports extensions that can add application features as if they were implemented in Flask itself. Flask is an excellent choice for building smaller application program interfaces and web services. Flask lets us focus on what the users are requesting and what kind of response to give back.

In the existent solution, the signals produced by the generator are directly transmitted to the deep brain without any control. These signals are vulnerable to external attacks by intruders or adversaries. In our solution, we propose the use of a separate classifier to detect the presence of the attack within genuine signals.

4.2 Client: stimulation signal generator

The solution involves one client (stimulation signal emulator), one server, and one remote IoT for reporting the results. We make both of them available on the web for public or privet users. This client is in the continuous link (client/server socket) with the server (stimulation classifier) for the transmission of the stimulation signals or attack patterns. The client produces emulated genuine signals and attack patterns to play the role of the deep brain stimulator. It assumes the potential presence of an adversary near the stimulator, introducing several attack patterns. On the server side, a receiving process is in continuous listening to the client, which reads the received signals and passes them through a classifier based on DNN to check whether the signal is an attack or a genuine signal.

The flowchart of the proposed system architecture is shown in Fig. 7.4. In the client, we define the parameters adjusting the probability value or the percentage of attacked signals in the flow. The random process of selecting the attack pattern is also called once the client decides to send an attack instead of a genuine signal. Once the stimulation signal is generated, a client/server socket routine transmits the signal to the server periodically for classification. Note here, that the role of the client is limited to the generation of the stimulation signal and the attack pattern only. The DNN predictor (deep classifier) is implemented on the server only. The prediction is done blindly without having any prior idea on the statistics of the stimulation signals produced by the client.

The web application including the signal generation and the web interface available for remote browsers is running on the client. Fig. 7.3 shows this web interface, which enables the personalization of the IP address and the port number of the server running the DNN model. In addition, the probability of attack (in %) can be personalized through this web interface. Pushing on one button specific to one attack pattern or a mixture of attack patterns is enough to validate the form and start the generation of the emulated signals.

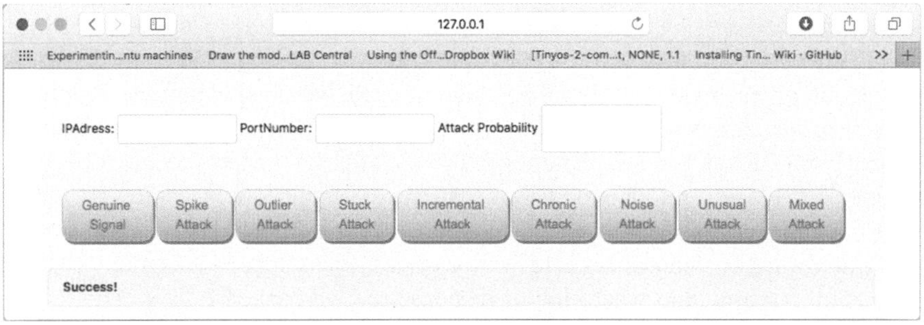

FIGURE 7.3

The web interface on the client side.

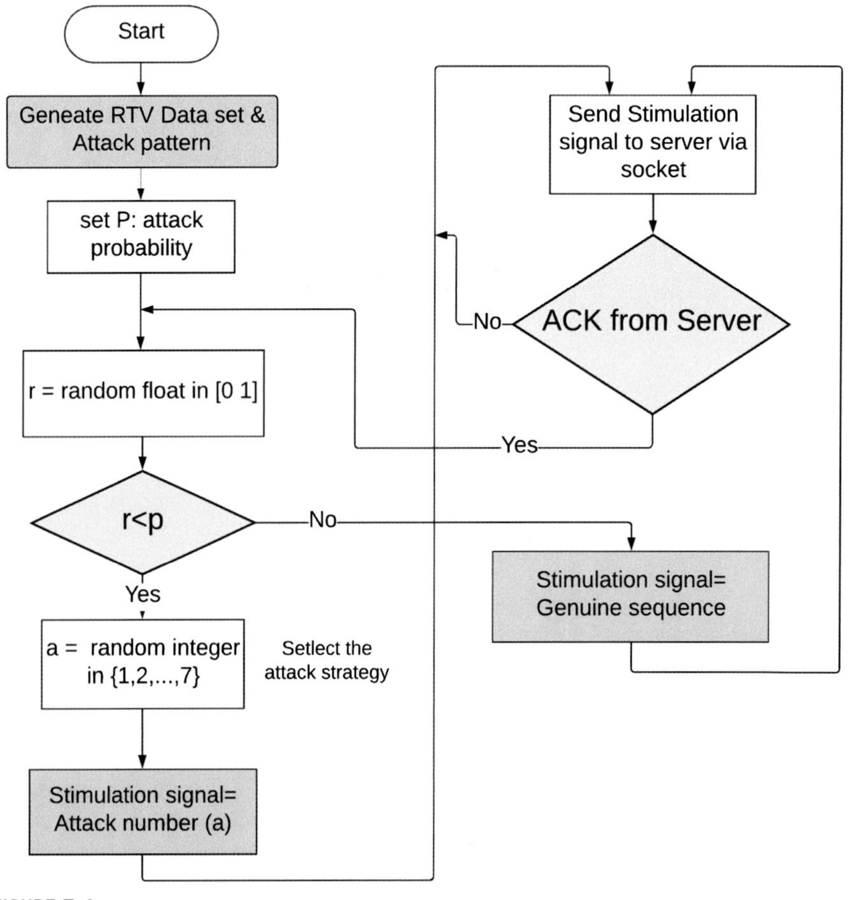

FIGURE 7.4

Flowchart of the client's algorithm.

4.3 Server: deep neural network classifier

The deep predictor running on the server, detailed in the previous section, ensures the classification of the signals received from the client. Hereafter, we propose the details of the implementation of the server and the technical solutions and hardware we selected. The core of the server is the flexible Raspberry Pi3 with the following features [34]:

- Broadcom BCM2837B0, Cortex-A53 64-bit SoC @ 1.4 GHz
- BCM43438 wireless LAN and Bluetooth Low Energy (BLE) on board
- Memory 1 GB LF DDR2 SDRAM,
- On board storage: Micro SDHC Slot for up to 32 GB,
- On board network:10/100 Mbits/s Ethernet port,
- Four USB 2.0 ports,
- Power source 5 V via Micro USB or GPIO header.

The processing power of the Raspberry Pi3 is suited for the web application on the server. This web application is continuously running and supporting the display of the results on remote browsers, the client/server communication to receive stimulation signals from the client, and the processing module for DNN classifier. The flowchart of the data processing on the server is given by Figs. 7.5–7.7.

On the client side, the web application is running using python, JavaScript, and HTML5 to give parameters for the web application to start the communication with the server and the generation of the stimulation signals and one or a mixture of the attack patterns.

In the existent solution, the signals produced by the pulse generator are directly transmitted to the deep brain without any control and they are vulnerable to external attacks. In our solution, we propose the use of separate DNN classifier to detect the presence of an attack within the genuine signals. The defender module, suited for the classification of the signals, is able to block any attack signal before being transmitted to the electrodes implanted in the brain. This device is designed on a Raspberry Pi3, running continuously.

The operating system controlling the Raspberry Pi3 is a modified version of Debian Linux optimized for the advanced RISC Machines (ARM v8-A) Cortex -A 53 architecture. The raspi-config command is used to configure the operating system.

The main goal of the proposed attack predictor is to use DNN to detect intrusions and display the results on remote browsers via the web application, in addition to posting the results to an IoT server for remote users. On the server, two main applications are implemented. The first one supports the process that receives the data from the client and the protocols to manage the message exchanges between the server and the IoT. The third one ensures the display of the results on a web interface to show the results in live via a web browser.

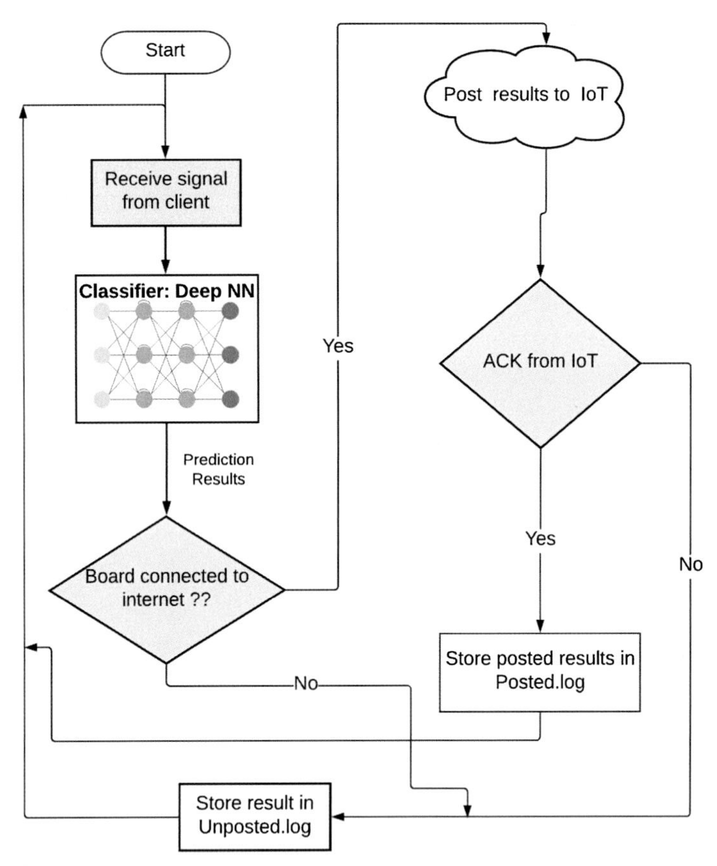

FIGURE 7.5

Flowchart of the data processing in the server.

5. Performance evaluation

Our system consists of (1) one data emulator, designed for the generation of the stimulation signal and attack patterns, and (2) one processing node (server) DNN classifier receiving the emulated data as input and produces, as output, the prediction of the presence of an attack or not. The hardware used for the emulator block is a Dell PC with Ubuntu 16.04, 4 GB RAM. The processing node, supporting the deep classifier, consists of a Raspberry Pi3 board and a Debian-based Linux operating system with Python 2.7-based codes. Once the signal is classified, the response "fake measurement" or "genuine measurement" is transmitted to an IoT or a server, continuously displaying the results for doctors or nurses supervising the patient. Genuine measurement is transmitted to the electrodes of the deep brain and the attacked signals are withdrawn.

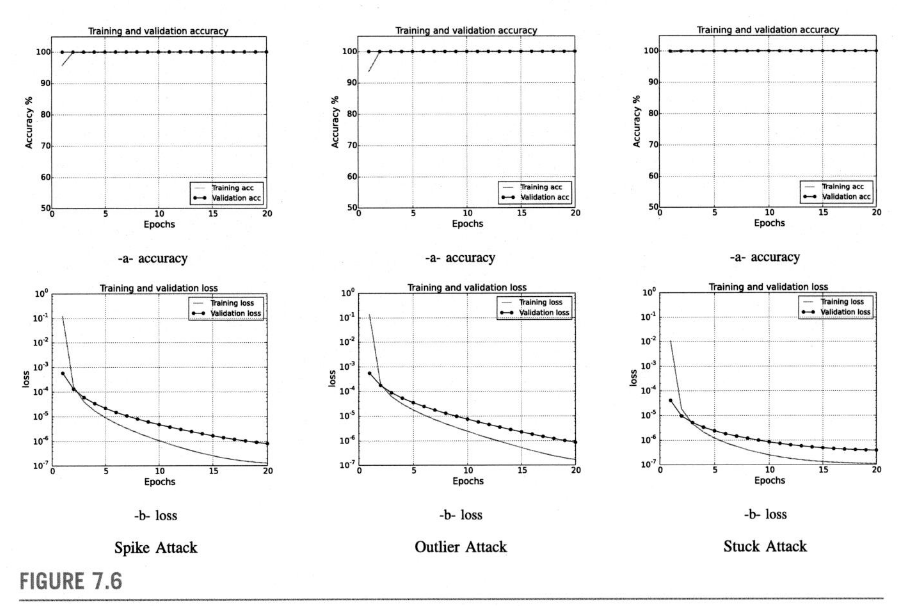

FIGURE 7.6

Accuracy and loss performances for Spike, Stuck, and Outlier attacks.

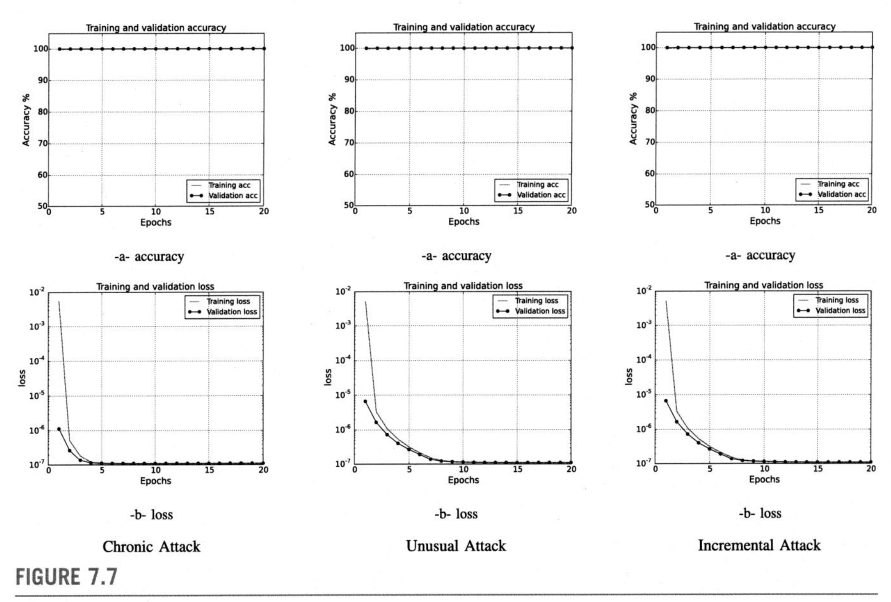

FIGURE 7.7

Accuracy and loss performances for Chronic, Incremental and Outlier attacks.

For DBS, the rest tremor velocity (RTV) is used for the training of the deep-learning model. Usually, RTV has a frequency of about 46 Hz for controlled neurological patients [35]. RTV is also the main variable to be measured for DBS-implanted patients. RTV is defined as the tremor that occurs when the muscles are not being voluntarily moved. The recordings in this database are the RTV in index finger. The dataset used in this study is composed of files from 30 patients, who were diagnosed with tremor dominant, and who all underwent surgery for the implantation of a neurostimulator (DBS treatment) at the John Radcliffe Hospital in Oxford, UK. The local research ethics committee of the Oxfordshire Health Authority approved the recordings, and informed consent was obtained from each patient.

To improve the security of the brain stimulator, in the following, we evaluate the performance of the deep classifier designed to predict the different attack patterns. Indeed, we evaluated the accuracy and the training loss that is the mean square error between the true and the predicted signal pattern. As described in Ref. [36], seven attack patterns, Spike, Outlier, Stuck, Incremental, Chronic, Noise, and Unusual are employed during the training of deep learning. For the prediction phase, we evaluated the positive predictive value, the negative predictive value, the specificity, and the sensitivity described in Section 3. The number of times the events "true positive," the "false positive," "true negative," and "false negative" are counted at each experience. "*true positive*": designs the event when the classifier succeeds predicting "fake measurement." However, "*false positive*" is the event when the classifier fails to predict a "genuine measurement." "*True negative*" is the event when the classifier succeeds to predict "genuine measurement." However, "*false negative*" is the event when the classifier fails to predict "fake measurement."

The deep training is designed using Tensor flow and Keras libraries [37,38]. The datasets are composed of 1300 genuine and 1300 attacked sequences collected from real measurements. Each sequence contains 300 samples and one label indicating if the sequence is genuine (label 0) or attacked (for example, Spike attack has label 1, Outlier has label 2 and so on ... Data augmentation is ensured by applying simple and complex transformation on the dataset. To increase the dataset size, we applied random rotations and we added a noise to each sequence. Augmentation can help to overcome the increasingly large requirements of deep-learning models. For the training, we evaluated the performance of the system by including one specific attack pattern, and we see how our classifier is able to detect attacked sequences from a flow of received sequences.

In Figs. 7.8 and 7.9, we present the results of the evaluation of the deep classifier. As we can see, the accuracy is more than 97%. Since we parameterized the emulator to produce 30% attacks as opposed to genuine measurements, the attacks predicted by the deep classifier are around 0.27.

The positive predictive value is close to 100%. This means that the classifier detects an attack in the sole case, where the attack is really existing and transmitted by the emulator. This means the number of times the classifier predicts an attack when a genuine signal was produced by the emulator is close to zero. The negative predictive value is around 98%. This means that sometimes, the classifier makes errors by

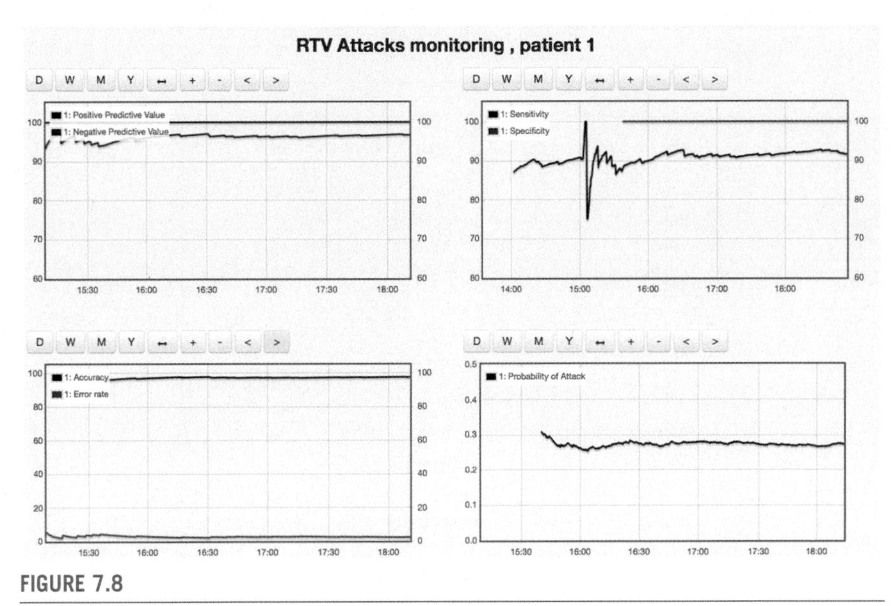

FIGURE 7.8

Performance analysis of the classifier with respect to the accuracy, positive predictive value, negative predictive value, Error rate, Probability of Attack.

predicting genuine signals when the emulator produces an attack, and this is very dangerous for the patient, as some attacks are not detected. The specificity is close to 93%, meaning that some prediction error occurs when the classifier predicts an attack and, in fact, there is no attack. This is similar to a false alarm. Specificity is the fraction of those without attack who will have a negative prediction result. Sensitivity is the probability that a prediction will indicate an "attack" among those with the attack.

6. Conclusion

To enhance the security of the signals that emerged in deep brain stimulators, we designed a monitoring system, based on deep learning on Raspberry Pi3. We show that deep learning presents an accuracy close to 97%, in learning and in predicting fake signals. We prove also that Raspberry Pi3 is able to support a web application including the deep classifier (prediction) and the web interface, in addition to the web engine (Flask). In future work, we can focus our efforts on the design of a low complexity monitoring system to reduce energy consumption and on the design of a complete monitoring system working on three modes. Learning mode, upgrade mode, and monitoring mode able to make learning from stored sequences and update to the learning model.

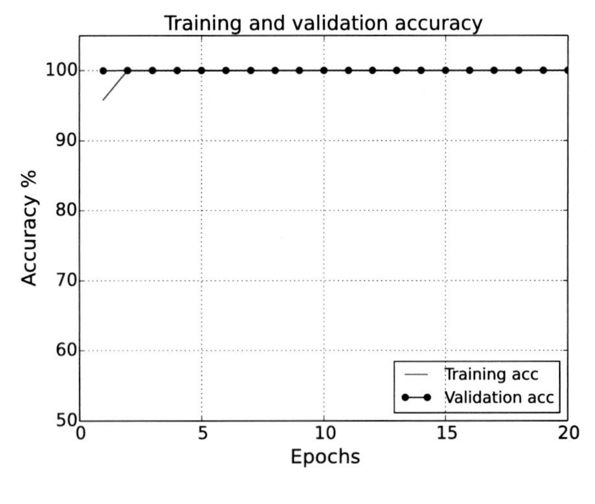

-a- accuracy

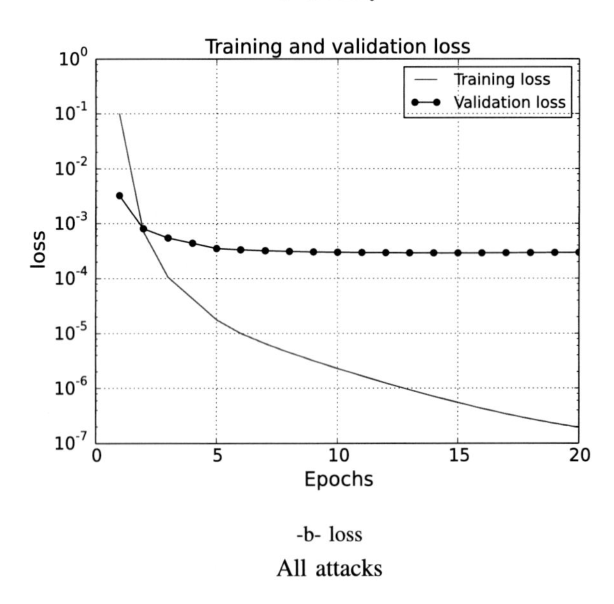

-b- loss

All attacks

FIGURE 7.9

Accuracy and loss performances for deep learning system involving all attacks.

Acknowledgments

This work was made possible by the National Priority Research Program (NPRP) award [NPRP8-408-2-172] from the Qatar National Research Fund (QNRF): a member of the Qatar Foundation. The work of Amr Mohamed and Mohsen Guizani is also supported in part by Qatar University Grant QUHI-CENG-19/20-1. The findings achieved herein are solely the responsibility of the authors.

References

[1] Long D. Electrical stimulation of the nervous system for pain control. Electroencephalography & Clinical Neurophysiology – Supplement 1978;34:343–8.

[2] Miocinovic S, Somayajula S, Chitnis S, Vitek JL. History, applications, and mechanisms of deep brain stimulation. JAMA Neurology 2013;70(2):163–71.

[3] Ostrem JL, Starr PA. Treatment of dystonia with deep brain stimulation. Neurotherapeutics 2008;5(2):320–30.

[4] Okun MS, Zeilman PR. Parkinson's disease: deep brain stimulation a practical guide for patients and families. Incorporated: National Parkinson Foundation; 2014.

[5] Priori A, Foffani G, Rossi L, Marceglia S. Adaptive deep brain stimulation (adbs) controlled by local field potential oscillations. Experimental Neurology 2013;245: 77–86.

[6] Swann NC, de Hemptinne C, Thompson MC, Miocinovic S, Miller AM, Ostrem JL, Chizeck HJ, Starr PA, et al. Adaptive deep brain stimulation for Parkinson?s disease using motor cortex sensing. Journal of Neural Engineering 2018;15(4):046006.

[7] Herron J, Chizeck HJ. Prototype closed-loop deep brain stimulation systems inspired by norbert wiener. In: IEEE conference on Norbert Wiener in the 21st century (21CW), vol. 2014. IEEE; 2014. p. 1–6.

[8] Choi W, Lee Y, Lee D, Kim H, Park JH, Kim IS, Lee DH. Less communication: energy-efficient key exchange for securing implantable medical devicesvol. 2018. *S*ecurity and Communication Networks; 2018.

[9] Bu L, Karpovsky MG, Kinsy MA. Bulwark: securing implantable medical devices communication channels. Computers & Security; 2018.

[10] Rushanan M, Rubin AD, Kune DF, Swanson CM. Sok: security and privacy in implantable medical devices and body area networks. In: 2014 IEEE symposium on security and privacy. IEEE; 2014. p. 524–39.

[11] Ankarali ZE, Abbasi QH, Demir AF, Serpedin E, Qaraqe K, Arslan H. A comparative review on the wireless implantable medical devices privacy and security. In: 2014 4th International conference on wireless mobile communication and healthcare-transforming healthcare through Innovations in mobile and wireless technologies (MOBIHEALTH). IEEE; 2014. p. 246–9.

[12] Li C, Raghunathan A, Jha NK. Hijacking an insulin pump: security attacks and defenses for a diabetes therapy system. In: 2011 IEEE 13th International conference on e-health networking, applications and services. IEEE; 2011. p. 150–6.

[13] Paul N, Kohno T, Klonoff DC. A review of the security of insulin pump infusion systems. Journal of Diabetes Science and technology 2011;5(6):1557–62.

[14] Camara C, Peris-Lopez P, Tapiador JE. Security and privacy issues in implantable medical devices: a comprehensive survey. Journal of Biomedical Informatics 2015;55: 272–89.

[15] Wu F, Eagles S. Cybersecurity for medical device manufacturers: ensuring safety and functionality. Biomedical Instrumentation & Technology 2016;50(1):23–34.

[16] Rathore H, Wenzel L, Al-Ali AK, Mohamed A, Du X, Guizani M. Multi-layer perceptron model on chip for secure diabetic treatment. IEEE Access 2018;6. 44 718–744 730.

[17] Rathore H, Al-Ali A, Mohamed A, Du X, Guizani M. Dlrt: deep learning approach for reliable diabetic treatment. In: GLOBECOM 2017-2017 IEEE global communications conference. IEEE; 2017. p. 1–6.

[18] Rathore H, Mohamed A, Al-Ali A, Du X, Guizani M. A review of security challenges, attacks and resolutions for wireless medical devices. In: 2017 13th International wireless communications and mobile computing conference (IWCMC). IEEE; 2017. p. 1495–501.
[19] Halperin D, Heydt-Benjamin TS, Ransford B, Clark SS, Defend B, Morgan W, Fu K, Kohno T, Maisel WH. Pacemakers and implantable cardiac defibrillators: software radio attacks and zero-power defenses. In: 2008 IEEE symposium on security and privacy (sp 2008). IEEE; 2008. p. 129–42.
[20] Malasri K, Wang L. Securing wireless implantable devices for healthcare: ideas and challenges. IEEE Communications Magazine 2009;47(7):74–80.
[21] Pournaghshband V, Sarrafzadeh M, Reiheret P. Wireless mobile communication and healthcare. 2013. p. 163–72.
[22] Ng H, Sim M, Tan C. Security issues of wireless sensor networks in healthcare applications. BT Technology Journal 2006;24(2):138–44.
[23] Hei X, Du X, Lin S, Lee I. Pipac: patient infusion pattern based access control scheme for wireless insulin pump system. In: 2013 Proceedings IEEE INFOCOM. IEEE; 2013. p. 3030–8.
[24] Leavitt N. Researchers fight to keep implanted medical devices safe from hackers. Computer 2010;43(8):11–4.
[25] Mozaffari-Kermani M, Sur-Kolay S, Raghunathan A, Jha NK. Systematic poisoning attacks on and defenses for machine learning in healthcare. IEEE Journal of Biomedical and Health Informatics 2014;19(6):1893–905.
[26] Pycroft L, Boccard SG, Owen SL, Stein JF, Fitzgerald JJ, Green AL, Aziz TZ. Brainjacking: implant security issues in invasive neuromodulation. World Neurosurgery 2016;92:454–62.
[27] Bengio Y, Courville A, Vincent P. Representation learning: a review and new perspectives. IEEE Transactions on Pattern Analysis and Machine Intelligence 2013; 35(8):1798–828.
[28] Marblestone AH, Wayne G, Kording KP. Toward an integration of deep learning and neuroscience. Frontiers in Computational Neuroscience 2016;10:94.
[29] Khalid S, Khalil T, Nasreen S. A survey of feature selection and feature extraction techniques in machine learning. In: 2014 science and information conference. IEEE; 2014. p. 372–8.
[30] Guyon I, Gunn S, Nikravesh M, Zadeh LA. Feature extraction: foundations and applications, vol. 207. Springer; 2008.
[31] Lai Y-K, Zhou Q-Y, Hu S-M, Wallner J, Pottmann H. Robust feature classification and editing. IEEE Transactions on Visualization and Computer Graphics 2006;13(1): 34–45.
[32] Le QV, Ngiam J, Coates A, Lahiri A, Prochnow B, Ng AY. On optimization methods for deep learning. In: Proceedings of the 28th International conference on machine learning. Omnipress; 2011. p. 265–72.
[33] Grinberg M. Flask web development: developing web applications with python. O'Reilly Media, Inc.; 2018.
[34] Raspberrypi. Raspberry pi 3 model B. https://www.raspberrypi.org/products/raspberry-pi-3-model-b/.
[35] Beuter A, Titcombe M, Richer F, Gross C, Guehl D. Effect of deep brain stimulation on amplitude and frequency characteristics of rest tremor in Parkinson?s disease. Thalamus & Related Systems 2001;1(3):203–11.

[36] Rathore H, Al-Ali A, Mohamed A, Du X, Guizani M. A novel deep learning strategy for classifying different attack patterns for deep brain implants. IEEE; 2019.
[37] Abadi M, Barham P, Chen J, Chen Z, Davis A, Dean J, Devin M, Ghemawat S, Irving G, Isard M, et al. Tensorflow: a system for large-scale machine learning. In: 12th {USENIX} symposium on operating systems design and implementation ({OSDI} 16); 2016. p. 265–83.
[38] Chollet F. Keras: Theano-based deep learning library. 2015. Code: https://github.com/fchollet.Documentation:http://keras.io.

CHAPTER

Blockchain applications for healthcare

8

Heena Rathore, PhD, BE [1], Amr Mohamed, PhD [2], Mohsen Guizani[2]

[1]*Department of Computer Science, University of Texas, San Antonio, TX, United States;* [2]*Professor, Department of Computer Science and Engineering, Qatar University, Doha, Qatar*

1. Introduction

Healthcare systems is an architectural paradigm coupled with pervasive sensing and communication technologies to provide multiple benefits to the medical environment. It is an engineered system, where the physical system or process is augmented with cyber components, such as computational hardware and communication network. These components are very tightly integrated with each other, which means functionality of one component is dependent on the other component. Key areas of research, while designing such systems to be smart, efficient and flexible, are stability, reliability, robustness, security, and privacy. However, rapid advancements in the enabling technologies, have also exposed such systems to serious and profound risks. If such risks are not managed, we would lose the incredible benefits that they can provide.

Blockchain is a secure digital ledger of transactions that can be configured to record, not only transactions in the financial world, but also in other areas, where maintaining a historical evidence of the transactions has value. It is the key technology behind bitcoins, a type of crypto currency. Bitcoins were developed after the financial crisis of 2009 as an alternative to traditional currency [1]. It is widely considered that one of the many reasons behind this crisis was the single point of failure, exemplified by how centralized banks maintained financial records. There was no oversight in this process and hence the lack of fault-tolerant checks and balances. Financial institutions, for a long time, have talked about the need for distributed decision-making process, but not acted on it, till the advent of crypto currencies fueled by the blockchain technology. A group of anonymous hackers, with alias of Satoshi Nakamoto, were responsible for writing the first set of codes. It is a type of mechanism, which validates, verifies, and confirms the transactions by recording them in a distributed ledger of blocks. It implements a consensus protocol to arrive at agreement on the validity of transaction by creating a chain of blocks. This immutable chain of blocks is trusted and verified, thereby making them a highly secure mechanism for maintaining a distributed ledger of transactions.

Initially, the blockchain technology was primarily utilized for protecting the financial transactions, smart contracts, storage systems, and notary. However, its

Energy Efficiency of Medical Devices and Healthcare Applications. https://doi.org/10.1016/B978-0-12-819045-6.00008-X

benefits were soon recognized by other applications such as supply chain, healthcare, transportation, and energy, as the industry realized that it can improve efficiency by adopting blockchain. This has spawned an active area of work, wherein researchers and scientists are now looking at other applications where this technology can be utilized. Energy, health, and transportation are some of the most commonly cited applications. In this chapter, we aim to

- Provide a holistic survey of applications where blockchain is being utilized in healthcare systems, both for record management and for securing implantable medical devices as shown in Fig. 8.1.
- Introduce IOTA as a light-weight alternative for distributed database system and compare it with Blockchain, to highlight its potential to reduce the processing requirements on IoT medical devices to enhance scalability.
- Provide a mathematical formulation to analyze when distributed ledger techniques such as blockchain are useful for an application.

Section 2 provides the details of applications of blockchain technology for healthcare systems. Section 3 gives the comparison of blockchain with respect to IOTA. Section 4 provides a mechanism to check if the application needs blockchain. Section 5 concludes the chapter.

2. Blockchain technology for HealthCare systems

Blockchain is now being used for record management in applications such as public health and medical research based on personal patient data. Evaluation metrics based on feasibility, intended capability, and compliance can be used to assess blockchain based decentralized applications in the area of healthcare [2]. The underlying benefit of blockchain, critical for health data, is that it is impossible to change

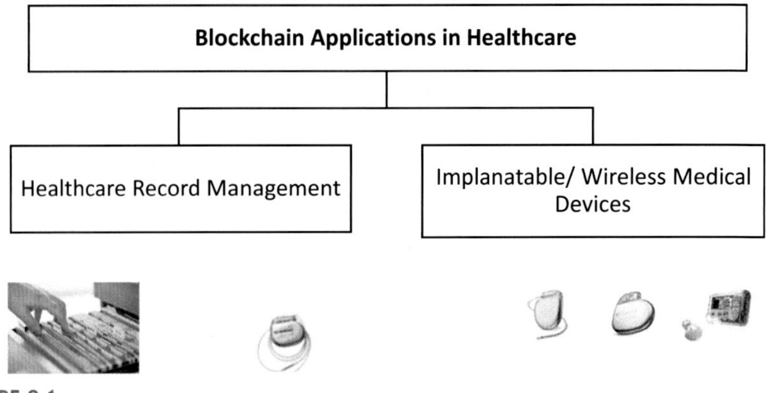

FIGURE 8.1

Blockchain applications in healthcare systems.

or delete a record without leaving a digital trace of the attempt to do so. Many countries, such as Estonia, are using blockchain to secure health and clinical trial records by linking access to data with permission settings. Blockchain also offers security through transparency, which enables the scanning of barcode-tagged drugs and tracking them through secure digital blocks whenever they change hands, thereby reducing the chances of counterfeiting. This can be further secured by allowing only authorized parties at the far end of the supply chain to access the records in real-time. There are diverse set of applications where blockchain can be utilized, namely data sharing, access control, health records, managing an audit trail, supply chain [3]. Active scientific work being done in some of these areas is surveyed next.

2.1 Healthcare record management

The management of integrity of the healthcare records and clinical preliminaries is pivotal. From the time instance when a medical record is created and marked, administrators are required to maintain evidence that the same has not been modified illegally, thereby maintaining the sanctity of the record. The field of healthcare record management deals with interoperability, information exchange, and analytics. Healthcare interoperability, as defined in Ref. [4], can either be institution driven or patient driven. The shift toward the latter brings with it various difficulties related to patient consent governance, security, protection, and patient commitment. Many scientific papers have been published showing how blockchain can facilitate administration of digital access rules related to information aggregation, information availability, and liquidity, as presented in Ref. [6]. Furthermore, it also helps with the understanding of patient attribute and its immutable nature. Scientific work on how a patient could safely collaborate with numerous stakeholders, recognize themselves over every entity, and aggregate the health information using an abnormal structure in a persistent form has been presented in Ref. [5]. A study of how interoperability is addressed among healthcare blockchain applications can be found in Ref. [7]. The MedRec [8] model gives a proof-of-framework, which enables standards of decentralization and blockchain designs to anchor and interoperate across medical record systems. It uses the Ethereum smart contracts to organize the framework and generates a log, which oversees medical records while giving patients the ability to survey complete records, audit care records, and share information. In this work, an inventive method for coordinating with suppliers' current framework, organizing open APIs and system structure transparency, has also been presented. A method adapted to handle big personal health data using a tree-based approach has been presented in Ref. [9]. To provide protection to cloud-hosted records, an initiative using blockchain technology has been presented in Ref. [10]. BlockHIE [11] is a another blockchain based platform for healthcare data which combines off-chain storage and on-chain verification to provide privacy and authentication to records. An architectural design, presented in Ref. [12], uses blockchain to facilitate healthcare data sharing in a private and audit-ready manner. It also handles permission-based healthcare data access using blockchain. Furthermore, a centralized source of trust in favor of network consensus and prediction of consensus of proof of structural and semantic interoperability has been presented in Ref. [13]. Finally, in the area of healthcare analytics, a blockchain-based application for storing and managing the database of patients and doctors during surgery is stated in Ref. [14]. A framework, based on artificial systems, computational experiments, and parallel execution using blockchain technology, which brings in the benefits of a parallel healthcare system has been proposed in Ref. [15]. A proposal to manage health data at individual and institution level using private blockchain solution has been made in Ref. [16].

2.2 Implantable medical device security

Smart systems allow persistent remote patient monitoring, thereby making it a prominent healthcare technology [17,27–31] and presenting healthcare information as a valuable source of medical intelligence. In the previous section, we looked at research centered around management of medical information. In this section, we will study recent advances in the sharing of healthcare information which has the potential to significantly benefit the quality of healthcare data. A recent analysis from U.S. federal government sources suggests that $3.65 trillion was spent on health care in 2018 by US population; an amount larger than the Gross Domestic Product (GDP) of countries such as Brazil, the U.K., Mexico, Spain, and Canada. This amounted to an increase of 4.4% over 2017 [18]. Likewise, a report in the journal Health Affairs estimates an average annual growth rate of 5.5% from 2018 to 2027, which means that at this rate, by 2027, health care will be 19.4% of the country's entire GDP. National well-being consumption is anticipated to grow at a normal rate of 6.3% every year through 2019, thereby achieving 19.6% of GDP by 2019 [19]. By the end of the year 2019, the market share for medical devices is expected to reach $186 billion, thereby making it one of the biggest markets in this space [19]. U.S. exports in medical devices, as recognized by the Department of Commerce (DoC), surpassed $44 billion in 2015 [20]; largely fueled by major innovation happening in more than 6500 medical device companies in the United States.

A case study of blockchain and IoT powered healthcare is presented in Ref. [21]. An amalgamation of blockchain and IoT in healthcare can be used for collecting and processing data in real time and providing secure access and data exchange between care providers. Using a private blockchain based on the Ethereum protocol, authors in Ref. [22] have created a smart contract, where the sensors communicate with smart devices. Such contracts manage records of all events on the blockchain to support real-time patient monitoring and medical interventions. MeDShare [23] is another trust-less system to share medical data using cloud service providers via blockchain. It has been shown to achieve data provenance and auditing while sharing medical data with diverse set of entities. Authors in Ref. [9] have designed a decentralized, permissioned blockchain for user-centric health data sharing. It has been designed to protect privacy using channel formation scheme and enhance the identity management using the membership service in mobile healthcare applications. A system that ensures security of patient's data through self-management, thereby preventing privacy violation, has been presented in Ref. [24].

Table 8.1 presents the various blockchain use cases, design challenges and future directions in healthcare.

3. IOTA technology versus blockchain

As demonstrated by the expanse of the scientific research surveyed in this paper, blockchain technology has been gaining rapid popularity in the recent years. It has the potential for changing the way in which people work and communicate by laying the foundation for emerging applications using connected devices. However, it has certain limitations such as:

Table 8.1 Blockchain applications in healthcare systems.

Application domain	Objectives/Use-cases	Future directions
Healthcare interoperability [4,5]	Data exchange, interpretation and usage	Advanced data analytics, and supported by robust care and coordination
MedRec record management	Governs medical record access while providing patients with comprehensive record review, care auditability and data sharing	Gather custom integration requirement to build open standard
BlockHie [11]	Healthcare information exchange for electronic medical records and personal healthcare data	Off-chain storage and on-chain verification for privacy and authentication
Healthcare analytics [14]	Acquisition, storage, and sharing of health data	Blockchain with artificial intelligence for healthcare analytics
Blockchain and IoT powered [21]	Big data incorporation for data mining	Require consensus model, less computational costs for mining blocks and validating transactions
MedShare [23]	Data sharing model between cloud service providers	Decrease latency contributing to the processing and anonymization of data
Data sharing and privacy [9]	A tree-based data processing based on Hyperledger fabric and batching method to personal health data	Combing both personal health data and medical data together
Privacy Violation [24]	Anonymization, communication, data backup and recovery	Secure raw data rather than anonymizing

- It does not scale with number of connected devices as it is limited by its usage of block size and time needed for hash calculations.
- In many cases, transaction validation through consensus consumes high energy, particularly in the case of proof-of-work [2].
- In some cases, it mandates the need for transaction fees or some other reward mechanism for miners
- While it is not as centralized as the concept of a single bank, it is still dependent on a handful of big entities, such as miners
- The computational and storage requirements of the blockchain participants are extensive since they have to store the entire ledger, and participate in the transaction verification process as endorsers, or miners.

Due to these limitations, experts are questioning whether blockchain is truly a viable technology as the number of connected medical devices increases. Researchers are now looking at IOTA, a technology introduced in 2017, for transaction validation and security for the Internet of Things related applications [25]. It is better suited to meet IoT requirements such as low resource consumption, widespread interoperability, billions of nano-transactions and data integrity, as it is faster, energy, resource efficient, and quantum proof. IOTA is a progressive transactional system and information exchange layer, which has been designed for securing applications around the internet of things. It is based on directed acyclic graph called Tangle, which is a type data structure technique. It has been designed to overcome some of the inefficiencies of Blockchain. In IOTA network, each transaction needs to validate two previous transactions by conducting a modified proof of work that considers energy-efficient transaction validation. The underlying theory is that the network will scale faster as more transactions are being validated in parallel. IOTA has properties of scalability, resource and energy optimization, data transfer security, and quantum readiness. As shown in Fig. 8.2, in a directed acyclic graph

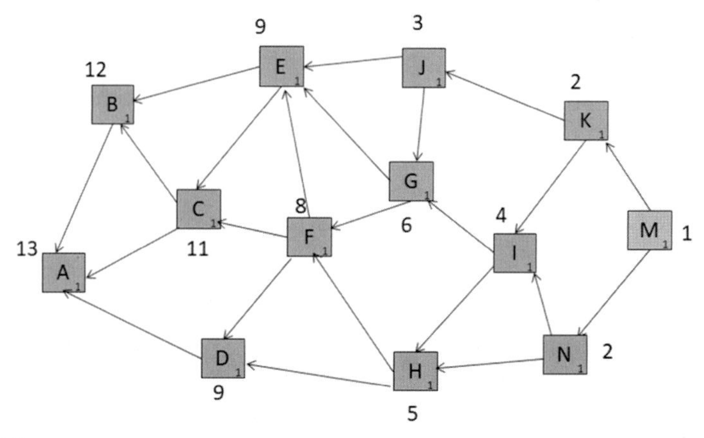

FIGURE 8.2

IOTA Tangle Explanation with Cumulative weight of the sites.

boxes are the sites (or transactions) The edges represent the link connecting the transactions which validate prior transactions. In order for a new transaction to be considered as a part of the network, the issuer has to verify two unconfirmed transactions, referred to as tips for the tangle. In addition, the new transactions have to perform three basic functions in order to be considered in the network namely:

- It must sign its transaction to authenticate itself in the network
- It must randomly select two, non-conflicting, transactions to validate
- It must work to validate these selected transactions

The direct weight of a node is generally expressed as three raised to a power of any real number. On the other hand, cumulative weight is obtained by adding the direct weight of all the sites, which has directly or indirectly verified the previous site. IOTA is being presented as a third-generation crypto currency that does not have the overhead of transaction validation but is nevertheless secure. Table 8.2 compares blockchain with IOTA.

IOTA is more decentralized compared to Blockchain. Blockchain will probably connect multiple IoT devices to one gateway and then the gateway would be a participant in the blockchain network, we can call this a clustered or a semi-decentralized approach. IOTA, on the other hand promotes the concept that the lightweight IoT device can be a participant in the IOTA network directly with reasonable processing, energy, and storage requirements, which allows any IoT devices to be integrated seamlessly into the IoTA network.

4. Does your application benefit from blockchain?

The requirements of many applications are adequately met by traditional relational databases; as such, they do not derive any major benifits by moving to a decentralized database such as Blockchain or IOTA. Hence, it is critical that one uses a decision support system (DSS) to determine if using a decentralized database adds sufficient value for the application, relative to the time and cost effort involved in

Table 8.2 Blockchain versus IOTA.

Blockchain	IOTA
Decentralized	Decentralized
Semi-distributed ownership	Truly distributed ownership
High computation power requirement	Lightweight, low computation power requirement
Energy requirements high	Energy requirements low
Not scalable	Scalable
Miners take transaction fees	No transaction fees
Not quantum resistant	Quantum computing protection

doing so [26]. In this section, we will present a mathematical formulation for making such a determination. The output of this mathematical formulation is an overall score s, which is defined by the following equation:

$$s = \frac{\sum_{i=1}^{N} a_i w_i}{N} \tag{8.1}$$

s is the overall score,
w_i is the weight for the metric under consideration $0 \leq w \leq 1.0$
N is the number of metrics under consideration
a_i is the scaling factor for the metric under consideration? $0 \leq a \leq 1.0$

The value of the scaling factor a is generally a complex function of the specific metric m being considered. There are two techniques in which the value of a may be determined. First approach is to perform a complex mathematical analysis of the time and cost complexity of the metric as it is tied to the use of a decentralized database for an application. A second approach is experiential in nature and uses a survey-based research to gather inputs with a cross-section of participants. The objective of such a survey is to collect the respondent's views on the importance of a metric for their application. In this paper, we take the second approach. Our goal is to compare the benefits of using a decentralized database technique for an application, using the overall score defined in Eq. (8.1). Application A1 is for a database that maintains grades for university students, henceforth referred to as University Database. Application A2 is for a database that maintains patient health records in a hospital database. This application is henceforth referred to as Hospital Database. For the purpose of this discussion, we consider this assessment with respect to a set of five metrics as described below

- ***Multiple Writers***: This metric considers the probability whether there will be multiple writers to the database. In other words, it considers whether the records in the database be entered only by one writer or will multiple writers be involved in modifying the records.
- ***Rogue Untrustworthy Actors***: This metric considers the probability whether there will be rogue actors in the database who can disrupt trust in the system. In other words, it considers whether all the actors in the system can be considered trustworthy or should we be concerned with some rogue actors introducing untrustworthiness in the system.
- ***Scalability***: This metric considers the need for scalability in the system and how the database architecture can adapt to an increased number of nodes in the system, without incurring additional overhead, cost, or bandwidth constraints.
- ***Historical Transaction Ledger:*** This metric considers the need for maintaining a historical ledger of transactions. Such a historical ledger could be required for governance and policy requirements and is typically found in banks, insurance companies and other highly regulated industries.

- ***Security:*** This metric considers the level of security needed in the database. While security is paramount for all application, this metric considers the relative importance of security for one application over the other.

We conducted a survey where participants were asked to assign values between 1 and 10 for these metrics for the University database and the Hospital database. A score of one implies least importance (or least risk) and a score of 10 implies highest importance (or highest risk). The survey was sent to 105 participants. The values in Table 8.3 are the average of the scores that the participants entered for each combination.

Assuming that the weight for all the metrics is 1.0, if we plug the above values in Eq. (8.1), we get the overall score of 0.67 for the university database application, versus an overall score of 0.78 for the hospital database. Using this framework, we can determine that a distributed ledger based decentralized database technique is better suited for a database that maintains patient records in a hospital database versus university database of student scores.

While the mathematical model described above provides a simple numerical output for decision making, in some situations, a radar chart is a useful visual tool that they used to compare different applications. A radar chart is a plot that consists of a sequence of equiangular spokes, called radii, with each spoke representing one of the variables. The data length of a spoke is proportional to the magnitude of the variable for the data point relative to the maximum magnitude of the variable across all data points. A line is drawn connecting the data values for each spoke. This gives the plot a star-like appearance and the origin of one of the popular names for this plot. A radar chart is a useful visual tool to determine which observations are most similar, i.e., are there clusters of observations and identify if there are there outliers. It also allows us to explore hybrid mechanisms for database design, which combines the strengths of relational and distributed databases. Fig. 8.3 shows a template radar chart for the two applications we considered above.

5. Closing remarks

As seen through the vast expanse of scientific research, Blockchain technology presents numerous opportunities for healthcare; however, it is not fully mature today

Table 8.3 Does your Application need Blockchain?

Metric	A1	A2
Multiple writers	0.5	0.3
Rogue Untrustworthy actors	0.83	0.95
Scalability	0.69	0.87
Historical transaction ledger	0.73	0.88
Security	0.69	0.9

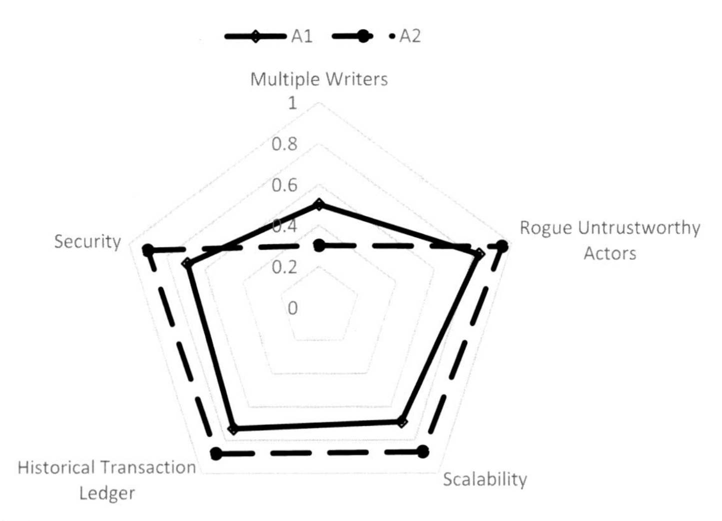

FIGURE 8.3

Radar chart: A visual tool for comparing the effectiveness of using a decentralized database for two different applications A1 and A2. A1 is the University Database and A2 is the Hospital database.

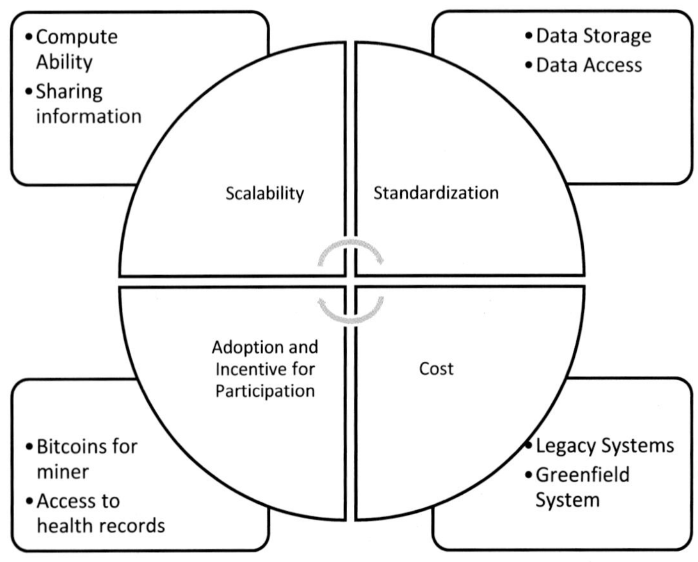

FIGURE 8.4

Challenges related to adoption of blockchain technology.

nor a panacea that can be immediately applied. Several technical, organizational, and behavioral economics challenges must be addressed before a healthcare blockchain can be adopted by organizations nationwide. In this section, we explore some of these challenges Fig. 8.4.

The first challenge is related to scalability constraints, which require us to manage the tradeoffs between transaction volumes and available computing power. There are two possible implementations of Blockchains, permissioned and permission-less. The latter are attractive as they enable broader access, are flexible and leverage computing power across the network. However, existing permissionless Blockchains, such as Ethereum or Bitcoin, face severe constraints as they can currently process approximately seven transactions per second, and with very high energy consumption for transaction validation. It is critical that this technology evolves to meet the demands of over 10 million users and 200,000 daily transactions. Permissioned Blockchains, for their part, can expedite the transaction processing times, but they may face computing power constrains due to reduced participation in the network. Plus, they more and more look like the centralized database, which essentially defeats the purpose of decentralized database.

The second challenge is related to data standardization and scope. In addition to evaluating the two different implementations highlighted above, teams should evaluate whether the data is stored on or off the blockchain. Size of the information (data) stored is a key metric to decide this. Patient information codified medical history, and other well confined information could be stored on the blockchain itself. Any type of detailed medical records, such as X-ray/MRI images, scans, freeform doctor notes could be stored off the Blockchain, to address the scalability. Users may need some form of standardization and a programming interface to manage and access such diverse set of data distributed across the enterprise.

Next challenge is related to adoption and incentives for participation. From an operations perspective, users need a network of interconnected computers to provide the computing power needed. Miners can then be incentivized to perform the job of validation, and add their compute power to the network in the form of cryptocurrency. For permissioned blockchains, participation could be encouraged through financial incentives or access to blockchain data in exchange for processing transactions. In addition to incentives for blockchain to work technically, further support may be needed to encourage users to adopt the technology and participate in a shared network.

The last and final challenge is related to costs of operating blockchain technology. While blockchain technology enables faster, near-real time transactions, the cost of operating such a system are not yet known. Health and government organizations have invested significantly in legacy systems, either in setup/configuration or long-term support and maintenance. Blockchain's open-source technology, properties, and distributed nature can help reduce the cost of these operations. Once a blockchain and its smart contracts are configured, the parameters become absolute, negating the need for frequent updates and troubleshooting. Since blockchain records are also immutable and stored across all participating users, recovery contingencies are unnecessary. Moreover, blockchain's transparent information structure could abolish many data exchange integration points and time-consuming reporting activities. The cost of computing power is derived from the volume and size of transactions submitted through the network; further varying by the type of transactions occurring on the chain (e.g., data storage vs. value exchange). Beyond the Bitcoin blockchain, there are scarce blockchains in full production, and as such, it is difficult to forecast the possible costs of operating a blockchain at scale within a private enterprise or among a consortium of partners. Therefore, to understand the potential costs of a fully scaled blockchain, customized to meet HHS and partner needs, targeted experiments and common blockchain guidelines are needed to iteratively test the technology with a view to scale.

6. Conclusion

A survey of the research papers done in this paper shows blockchain and their inherent combination of consensus algorithms, distributed data storage, and secure protocols can be utilized to build robustness and reliability in healthcare systems. Blockchain provides a distributed ledger system that validates transactions and ensures that no one can tamper with them, once they are added to the chain. If someone does try to tamper with the records, it is easy to spot the change in pattern and take defensive measures. Blockchain also eliminates the risks associated with a centralized architecture as the role of validating transactions is distributed with actors who have significant investment in the network, either measured as proof of stake or proof of investment. This emerging technology provides internet users the capability to create value and authenticate the digital information. It has the capability to revolutionize a diverse set of business applications ranging from sharing economy to data management and prediction markets. The survey results presented in this paper demonstrate that Blockchain has distinct advantages for healthcare applications as compared to other applications. The benefits of blockchain can be further amplified by using a light-weight distributed ledger system like IOTA.

Acknowledgment

This work was made possible by Qatar University Grant QUHI-CENG-19/20-1. The findings achieved herein are solely the responsibility of the authors.

References

[1] Markham JW. A financial history of the United States: from Enron-era scandals to the subprime crisis (2004–2006); from the subprime crisis to the Great Recession (2006–2009). Routledge; 2015.

[2] Zhang P, Walker MA, White J, Schmidt DC, Lenz G. Metrics for assessing blockchain-based healthcare decentralized apps. In: 2017 IEEE 19th International conference on e-health networking, applications and services (Healthcom). IEEE; October 2017. p. 1–4.

[3] Hölbl M, Kompara M, Kamišalić A, Nemec Zlatolas L. A systematic review of the use of blockchain in healthcare. Symmetry 2018;10(10):470.
[4] HIMSS, What is interoperability? [accessed on November 14, 2019].
[5] Gordon WJ, Catalini C. Blockchain technology for healthcare: facilitating the transition to patient-driven interoperability. Computational and Structural Biotechnology Journal 2018;16:224–30.
[6] Gökalp E, Gökalp MO, Çoban S, Eren PE. Analysing opportunities and challenges of integrated blockchain technologies in healthcare. In: EuroSymposium on systems analysis and design. Cham: Springer; September 2018. p. 174–83.
[7] Zhang P, White J, Schmidt DC, Lenz G. Applying software patterns to address interoperability in blockchain-based healthcare apps. arXiv preprint arXiv:1706.03700. 2017.
[8] Ekblaw A, Azaria A, Halamka JD, Lippman A. A Case Study for Blockchain in Healthcare:"MedRec" prototype for electronic health records and medical research data. In: Proceedings of IEEE open and big data conference, vol. 13; August 2016. p. 13.
[9] Liang X, Zhao J, Shetty S, Liu J, Li D. Integrating blockchain for data sharing and collaboration in mobile healthcare applications. In: 2017 IEEE 28th Annual International symposium on personal, indoor, and mobile radio communications (PIMRC). IEEE; October 2017. p. 1–5.
[10] Esposito C, De Santis A, Tortora G, Chang H, Choo KKR. Blockchain: a panacea for healthcare cloud-based data security and privacy? IEEE Cloud Computing 2018;5(1): 31–7.
[11] Jiang S, Cao J, Wu H, Yang Y, Ma M, He J. Blochie: a blockchain-based platform for healthcare information exchange. In: 2018 IEEE International conference on smart computing (SMARTCOMP). IEEE; June 2018. p. 49–56.
[12] Theodouli A, Arakliotis S, Moschou K, Votis K, Tzovaras D. On the design of a blockchain-based system to facilitate healthcare data sharing. In: 2018 17th IEEE International Conference on trust, security and privacy in computing and Communications/ 12th IEEE International Conference on big data science and engineering (TrustCom/ BigDataSE). IEEE; August 2018. p. 1374–9.
[13] Peterson K, Deeduvanu R, Kanjamala P, Boles K. A blockchain-based approach to health information exchange networks. In: Proc. NIST workshop blockchain healthcare, vol. 1; 2016. p. 1–10.
[14] Le Nguyen T. Blockchain in healthcare: a new technology benefit for both patients and doctors. In: 2018 Portland International Conference on management of engineering and technology (PICMET). IEEE; August 2018. p. 1–6.
[15] Wang S, Wang J, Wang X, Qiu T, Yuan Y, Ouyang L, Guo Y, Wang FY. Blockchain-powered parallel healthcare systems based on the ACP approach. IEEE Transactions on Computational Social Systems 2018;(99):1–9.
[16] Bhuiyan MZA, Zaman A, Wang T, Wang G, Tao H, Hassan MM. Blockchain and big data to transform the healthcare. In: Proceedings of the International Conference on data processing and applications. ACM; May 2018. p. 62–8.
[17] Rathore H, Mohamed A, Al-Ali A, Du X, Guizani M. A review of security challenges, attacks and resolutions for wireless medical devices. In: Wireless communications and mobile computing Conference (IWCMC), 2017 13th International. IEEE; June 2017. p. 1495–501.
[18] Fortune, U.S. Health care costs Skyrocketed to $3.65 trillion in 2018 [accessed on 14 Nov 2019].

[19] Medtech Switzerland, 2017, "The U.S. Market for medical devices: opportunities and challenges for Swiss companies".

[20] SelectUSA, 2017, https://www.selectusa.gov/medical-technology-industry-united-states [accessed on 11 Jan, 2017].

[21] Karafiloski E, Mishev A. Blockchain solutions for big data challenges: a literature review. In: IEEE EUROCON 2017 - 17th International Conference on smart technologies. IEEE; July 2017. p. 763−8.

[22] Griggs KN, Ossipova O, Kohlios CP, Baccarini AN, Howson EA, Hayajneh T. Healthcare blockchain system using smart contracts for secure automated remote patient monitoring. Journal of Medical Systems 2018;42(7):130.

[23] Xia QI, Sifah EB, Asamoah KO, Gao J, Du X, Guizani M. MeDShare: trust-less medical data sharing among cloud service providers via blockchain. IEEE Access 2017;5: 14757−67.

[24] Yue X, Wang H, Jin D, Li M, Jiang W. Healthcare data gateways: found healthcare intelligence on blockchain with novel privacy risk control. Journal of Medical Systems 2016;40(10):218.

[25] Popov S. The Tangle. White Paper 2018.

[26] NIS Cooperation Group, "EU coordinated risk assessment of the cybersecurity of 5G networks" [accessed on November 14, 2019].

[27] Rathore H, Fu C, Mohamed A, Al-Ali A, Du X, Guizani M, Yu Z. Multi-layer security scheme for implantable medical devices. Neural Computing & Applications 2018: 1−14.

[28] Rathore H, Al-Ali AK, Mohamed A, Du X, Guizani M. A novel deep learning strategy for classifying different attack patterns for deep brain implants. IEEE Access 2019;7: 24154−64.

[29] Rathore H, Wenzel L, Al-Ali AK, Mohamed A, Du X, Guizani M. Multi-layer perceptron model on chip for secure diabetic treatment. IEEE Access 2018;6:44718−30.

[30] Rathore H, Al-Ali A, Mohamed A, Du X, Guizani M. DTW based authentication for wireless medical device security. In: 2018 14th International Wireless communications & mobile computing Conference (IWCMC). IEEE; June 2018. p. 476−81.

[31] Rathore H, Al-Ali A, Mohamed A, Du X, Guizani M. DLRT: deep learning approach for reliable diabetic treatment. In: GLOBECOM 2017 - 2017 IEEE Global communications Conference. IEEE; December 2017. p. 1−6.

Index

'*Note*: Page numbers followed by "f" indicate figures and "t" indicate tables.'

I

K

L

M

N

O

P

Q

R

Printed in the United States
By Bookmasters